Teaching in the Built Environment

In *Teaching in the Built Environment: Creating Transformational Active Learning Experiences*, three master teachers reveal classroom-ready strategies for maximizing undergraduate learning in built environment disciplines. Drawing on the authors' years of experience as education researchers and faculty at one of the nation's top design and construction schools, this primer empowers instructors to implement high-impact teaching practices in any educational setting, from intimate seminars to "mega classes" of hundreds of students. Chapters on internships, study abroad, and field experiences equip faculty with teaching tools guaranteed to work on or off campus.

Just as importantly, *Teaching in the Built Environment: Creating Transformational Active Learning Experiences* identifies the teaching tactics that *don't* work. The authors' candid reflections on their own failed pedagogical experiments help instructors avoid confidence-shaking missteps and encourage them to turn teaching struggles into future successes. A thorough review of the latest education research provides theoretical context and empirical support for strategies direct from the authors' award-winning classrooms, studios, and labs.

A must-read for built environment educators, from first-time faculty to classroom veterans, *Teaching in the Built Environment: Creating Transformational Active Learning Experiences* inspires teaching that will resonate long past the semester's end.

C. Ben Farrow, PhD, is the Associate Dean for Academic Affairs and International Programs in Auburn University's College of Architecture, Design and Construction. Prior to his administrative role, Ben had fifteen years of experience in industry primarily as a practicing structural engineer and eleven years teaching in Auburn's McWhorter School of Building Science. He served as undergraduate program chair in Building Science for four years, held both the Aderholt Professorship and the Hunt Professorship within the School, and was the 2020–21 president of the Associated Schools of Construction (ASC) which represents over 170 construction management programs worldwide. He has multiple teaching awards collected at the School, College, and Regional levels. He holds a doctorate in adult education, and his primary research areas of interest include student engagement within construction education and recruitment of students into the built environment.

Eric M. Wetzel, PhD, is an Associate Professor in Auburn University's McWhorter School of Building Science. He is the recipient of multiple teaching and service awards from academia and industry, including the 2021 Atlanta Alumni Building Science Fund for Excellence in Service Award, 2020 Associated General Contractors (AGC) of Alabama 40 Under 40, and the 2019 Auburn University Alumni Undergraduate Teaching Excellence Award. Eric is currently serving on the American Council for Construction Education Standards Committee and is an active member in the Associated Schools of Construction. He holds a doctorate in construction management, and his primary research interests include active learning within construction education and robotics in construction. Outside of the university, Eric has over 12 years of experience as a construction consultant, specializing in project controls for the water and wastewater industry.

Tom M. Leathem, PhD, is the JE Wilborn Associate Professor in Auburn University's McWhorter School of Building Science. He is the Proceedings Editor for the Associated Schools of Construction and has held leadership roles with the organization since 2013. He serves on the American Council for Construction Education Training Committee and facilitates numerous professional development workshops in teaching and learning. He is the recipient of multiple teaching awards including the Associated Schools of Construction 2016 Excellence in Teaching Award, and the 2019 McWhorter School of Building Science Outstanding Faculty Award. He holds a doctorate in curriculum and assessment, and his main areas of research interest include construction education and assessment, collaborative education in the built environment, and innovation in construction education. Before entering the academic world, he worked in the commercial industrial construction industry for 11 years in various management roles.

Teaching in the Built Environment

Creating Transformational Active Learning Experiences

C. Ben Farrow, Eric M. Wetzel, and Tom M. Leathem

Foreword by George D. Kuh

Illustrations by David Smith

Routledge
Taylor & Francis Group
NEW YORK AND LONDON

Cover image: © Getty Images

First published 2022
by Routledge
605 Third Avenue, New York, NY 10158

and by Routledge
4 Park Square, Milton Park, Abingdon, Oxon, OX14 4RN

Routledge is an imprint of the Taylor & Francis Group, an informa business

Library of Congress Cataloging-in-Publication Data
Names: Wetzel, Eric, author. | Leathem, Tom, 1976– author. |
Farrow, C. Ben, author.
Title: Teaching in the built environment: creating transformational active learning experiences / Eric Wetzel, Thomas Leathem, C. Ben Farrow.
Description: New York, NY: Routledge, 2022. | Includes bibliographical references and index. |
Identifiers: LCCN 2021052023 (print) | LCCN 2021052024 (ebook) |
ISBN 9780367616786 (paperback) | ISBN 9780367616892 (hardback) |
ISBN 9781003106029 (ebook)
Subjects: LCSH: Active learning. | Transformative learning. |
College teaching—Methodology. | Education, Higher.
Classification: LCC LB1027.23 .W48 2022 (print) |
LCC LB1027.23 (ebook) | DDC 371.3—dc23/eng/20211213
LC record available at https://lccn.loc.gov/2021052023
LC ebook record available at https://lccn.loc.gov/2021052024

ISBN: 978-0-367-61689-2 (hbk)
ISBN: 978-0-367-61678-6 (pbk)
ISBN: 978-1-003-10602-9 (ebk)

DOI: 10.1201/9781003106029

Typeset in Goudy
by codeMantra

Contents

Figures

Foreword

George D. Kuh

There was a time when mentioning active or experiential learning at academic conferences conjured up images of students sitting in a classroom or outdoors, talking among themselves, with little of substance expected from the time spent. A cartoon with this message featured a teacher telling the class to break into groups of six and to return in twenty minutes ready to share what they discussed. The inference is that whatever students did was an afterthought to what really matters – the material covered in a lecture by an expert.

However, over the past couple of decades a fair amount of literature across a wide variety of disciplines has documented the value of putting students in situations where they are required to apply what they are learning to "real life" problems. Employers have called for more of these kinds of relevant, meaningful educational experiences so as to better prepare graduates to "hit the ground running" and function effectively in the workplace (Association of American Colleges and Universities, 2021). Among the various approaches favored by both employers and educational researchers are ensuring students participate in high-impact practices, or HIPs. Indeed, in a 2017 survey more the half of the responding provosts reported their institution was adding and scaling HIP opportunities so that more students would do one or more (https://www.learningoutcomesassessment.org/wp-content/uploads/2019/02/2018SurveyReport.pdf). Why this level of interest?

- Students who participate in one or more HIPs are more likely than their peers to complete a baccalaureate degree and report greater gains on a variety of desired learning and personal development outcomes.
- HIPs also are associated with salutary effects for students from historically underserved populations; that is, those first in their family to attend college, the academically underprepared, and those from racially minoritized populations who do at least one HIP tend to get a boost in their performance.
- Participating in a HIP also helps shrink the psychological size of the institution, fostering a greater sense of belonging thought to be a precursor to persistence and satisfaction with college.
- In addition, a HIP experience often puts students in unscripted situations that require them to apply what they know which helps deepen and integrate their learning.

- Finally, students who do multiple HIPs during college realize cumulative, additive effects in the areas just mentioned. In other words, generally the more HIPs the better.

Not every student acknowledges the benefits from HIPs participation to the extent inferred above. Nevertheless, the evidence so far suggests that few other undergraduate experiences, innovations or interventions yield so many benefits to participants, especially *when HIPs are done well* (Kuh, 2013).

The array of positive outcomes associated with HIP participation mentioned above is not surprising because they are composed of eight educationally effective features that are described later in this book and elsewhere (Kuh, 2008; Kuh, O'Donnell, & Rice, 2017). Standing alone, each of these features has a solid empirical foundation to qualify as a good practice in undergraduate education (Chickering & Gamson, 1987; Brownell & Swaner, 2008; Mayhew et al., 2016). When bundled together with other active learning activities as the authors of this book recommend, the impact is much greater.

The bulk of the HIP literature focuses on the characteristics of individual HIPs, such as service learning or an internship, as contrasted with the educationally effective features that make a HIP a HIP. This book is timely and instructive because it provides numerous concrete examples for *how to apply* and *integrate* these features with other effective educational practices for students to realize the potentially transformative effects of a high-quality undergraduate experience. Especially helpful is the range of applications, from traditional classrooms, labs and studios to technology-enhanced classes to intensive writing and intensive math courses, and capstone experiences.

Given the authors' disciplinary affiliations, the book understandably is pitched to colleagues in architecture, design, construction, and related fields. Even so, most of the concepts and examples are transferable to other applied areas of study as well liberal arts disciplines. Kuhdos [sic] to the authors who illustrate how to infuse active learning approaches which are the foundation of HIPs into any setting where learning and teaching occur, on and off the campus. All this merits a tip of the hat to Wetzel, Farrow, and Leathem for showing us why and how HIPs and other active and experiential learning approaches carefully designed and intentionally threaded together can foster high levels of learning and personal development.

References

Association of American Colleges and Universities. (2021). *How college contributes to workforce success: Employer views on what matters most*. Washington, DC: Association of American Colleges and Universities.

Brownell, J. E., & Swaner, L. E. (2008). *Five high-impact practices: Research on learning outcomes completion and quality*. Washington, DC: Association of American Colleges and Universities.

Chickering, A. W., & Gamson, Z. F. (1987). Seven principles for good practice in undergraduate education. *AAHE Bulletin, March*, 3–7.

Kuh, G. D. (2008). *High-impact educational practices: What they are, who has access to them, and why they matter.* Washington, DC: Association of American Colleges and Universities.

Kuh, G. D. (2013). Part 1: Taking HIPs to the next level. In G. D. Kuh, K. O'Donnell, and S. Reed (Eds.), *Ensuring quality and taking high-impact practices to scale* (pp. 1–14). Washington, DC: Association of American Colleges and Universities.

Kuh, G. D., O'Donnell, K., & Schneider, C. G. (2017). HIPs at ten. *Change: The Magazine of Higher Learning, 49*(5), 8–16.

Mayhew, M. J., Rockenbach, A. N., Bowman, N. A., Seifert, T. A., Wolniak, G. C., Pascarella, E. T., & Terenzini, P. T. (2016). *How college affects students, volume 3: 21st century evidence that higher education works.* San Francisco, CA: Jossey-Bass.

George D. Kuh is Chancellor's Professor Emeritus of Higher Education at Indiana University.

Acknowledgements

Our collective thanks go out to Steve Williams, George Kuh, and everyone who committed their time to reading the numerous iterations of the book and providing us feedback along the way. Without a doubt, the constructive and supportive feedback provided was instrumental in developing the book to what it is. Thank you to the numerous faculty who contributed their time to share stories of success and failures in their quest for teaching excellence. Your stories helped us develop the individual chapters and make the book richer. Thanks also goes to David Smith, our graphic designer, for not only putting up with the three of us trying to make decisions but also for your willingness to serve as a proofreader.

The authors collectively recognize the generations of outstanding teaching within the McWhorter School of Building Science at Auburn University and the motivation that work provided for this manuscript. Numerous faculty have created a culture of great teaching and learning, and any positives from this book are a direct result of their work.

We could not do an appropriate acknowledgment without recognizing each of our wonderful families who supported us in this journey.

Tom's Acknowledgment

To my wonderful wife Erin, thank you for your love and encouragement through many months of book meetings, weekend phone calls, and late-night text messages with my coauthors. To my son Brady and daughters Addison and Katelyn, thank you for your support and encouragement. You helped motivate me more than you can know. To my mom and dad, thank you for your endless encouragement that I can accomplish anything I put my mind to.

Ben's Acknowledgment

To my spouse, Julia, thank you for your love and endless, undeserved patience (and your proofreading skills). To my family including Abby, Joe, Meredith, and Gabe, thank you for supporting me during the arduous process of writing a book and encouraging me along the way. To my parents in heaven, thank you for inspiring me to always stretch for great teaching and learning in whatever role I served.

Eric's Acknowledgment

To my smart, beautiful, caring wife, Heather, thank you for everything. None of this is possible without you. To my kids, Colton and Annabeth, you bring your mom and I so much joy – we love you. To my parents, Ed and Veronica, and my brother Peter, thank you for a lifetime of support and love. To Tom's wife, Erin, and Ben's wife, Julia, I don't know how you do it... Just kidding. To my coauthors, Ben and Tom, I couldn't imagine writing a book with anyone else. Thank you both for an awesome experience.

1 A Call for Transformative Educational Experiences

Sarah is in her mid-30s and has been living in Omaha, Nebraska, for a few years. She likes it. Her job as an engineer for the city is fulfilling, and she feels confident in the work she is doing. Sarah is a proud alumna; the diploma in her office displaying her Bachelor of Science in Civil Engineering is hung where it couldn't be missed by a passerby. She has little knickknacks with the university logo and a sticker on her hardhat with the seal of her alma mater. It doesn't happen often, but occasionally, Sarah has an opportunity to lean back in her office chair and think about her time in college. She fondly recalls the football game in the blizzard, the party near the lake, her sorority bid day, and her study abroad experience in Bolivia.

The Bolivia trip was only a week, but what a week it was. They had spent much of the semester designing a water irrigation system for a remote village, an eight-hour drive from their initial destination, La Paz. Leading up to the trip, the group of ten students had really grown to trust each other. Their professor, Dr. Wake, spent a great deal of time preparing them for the experience and set clear expectations to create a workable irrigation system for this arid community. On arrival in the tiny village, Sarah was met by one of the local children who insisted she spend time with her coloring a picture. At a service to celebrate their arrival, Sarah heard several of the locals share (through a translator) what the project would mean to the community. As she prepared to start the week, it became very apparent the challenge that lay ahead. The altitude, the heat, and the available tools all seemed daunting to the task. However, she loved working side by side with the Bolivian locals and getting to learn more about each of them and their culture. Eduardo, who said very little but worked tirelessly, showed her some tricks to make the work easier. Over the course of the week, she could tell he was drawn to the students' passion for the work, and a connection was built that extended beyond languages. Before leaving, the villagers cooked a special dinner for the students, and each student was asked to share their experiences. Sarah vividly remembers the smell and taste of that meal, the passion in her voice as she shared her experience, the pride in completing the project, and the relationships she fostered with both students and villagers.

DOI: 10.1201/9781003106029-1

It has been 15 years since that trip, but Sarah's memories of Eduardo, the dinner, and the villagers feel recent. What was it that created such an impactful experience? It was only one week of her four-year college journey, and much of it was spent doing backbreaking work at altitude under a baking sun. Why did that experience stand out among all the educational experiences she had in college? It was a water irrigation system – who gets excited about that? What was Dr. Wake's role through the trip and was there something he did that could shed light on why this memory is so vivid? Perhaps a trip to Bolivia would be inherently memorable. But, manual labor for a week shouldn't yield such fond memories. What can we, as educators, learn from Sarah's experience – through her lens? Could the lasting impact of her trip be captured, and could a framework be developed that would maximize the opportunity for any student to have deep, impactful educational experiences that they would likely carry with them for the rest of their lives – an instilled passion? And finally, could this approach be scaled to cover most class types, so students could have at least one or two of these experiences each semester of their college career?

We, the authors, believe that by closely evaluating examples like Sarah's and others, we can answer all these questions and more. This book is our attempt at breaking down and harnessing these types of incredible learning experiences, in all different types of classes. A trip to Bolivia would surely create some excitement, but we will show that by being intentional in course design, it's possible to create that same passion in all Architecture, Engineering, and Construction (AEC) classes. By using a combination of existing pedagogical approaches, the literature behind High Impact Educational Practices (HIPs), impactful moments, and a novel approach to coursework design, anyone can create the environment to maximize the opportunity for a deeply impactful, lasting educational experience – a Transformational Active Learning Experience or if you prefer, a "TALE."

But why are TALEs necessary? The question isn't rhetorical. For decades, many universities have prioritized research over teaching. Bringing in external funding, publishing in journals, and writing books have been the primary concern of faculty, while teaching has been relegated at times to something done in between grant applications and graduate student meetings. Recently, due to external pressures (high cost of education, declining enrollment, rise of online education, etc.), we can see a shift; not to a point where grants and research become a miniscule factor of promotion and tenure, but perhaps weighted closer together. The shift has been trickling in for years with pedagogical foci on active learning and engagement, but with the onset of COVID-19, the rumblings from students and parents became boisterous yells. The obvious was exposed. To the students and their parents, the *experience* provided by colleges and universities is as important as the academic credentials.

Through student affairs, athletic departments, and alumni associations, many universities are well situated for the social experience of college, but we, the faculty and administrators, are the vanguard of the academic experience. Are we adding value to the student experience, or are we simply disseminating technical information, testing their knowledge, and granting a degree? It is easy for faculty to be critical of athletic budgets, recreation center pools, multimillion-dollar workout facilities, and gourmet dining in the student center, but if we are doing the bare minimum to engage students academically, then the complaints are hypocritical. Like it or not, universities must attract and retain students, and with the significant cost of higher education, we can no longer assume that our impressive research portfolios and lecture hall approach will bring them to campus. Football stadiums, dining halls, and dorm rooms may get students to apply, but teaching keeps them engaged, connected, and propels them toward a successful adult life. The most valuable faculty will be those who can both motivate the content and drive the experience. The most valuable administrators will be those who recognize the importance of the student experience and leverage resources towards improving it to create a competitive advantage for their university. Said another way, universities that differentiate themselves in the years ahead will not only convey knowledge but also curate learning experiences. The emphasis on academic student engagement must be a top priority.

At the time of writing this book, COVID-19 has impacted the traditional higher education landscape where students come together to live and study in close proximity to one another. Universities have rushed to maximize online learning while fortifying IT support structures. Yet, many of us have realized that dependence on community and in-person interaction are basic human needs. These dependences provide a platform on which the in-person student experience can be enhanced to elevate learning and to differentiate higher education programs.

In the subsequent chapters, we will discuss how to maximize your students' opportunity for a TALE. In order to understand our approach, we first provide some background in existing pedagogies and literature. Chapters 2 and 3 explain, in significant detail, how we use elements of HIPs and the science behind "Moments" – two important aspects of TALEs. Chapter 4 will show how these concepts, aligned with active learning strategies, can be structured to analyze the student learning experience. In Chapter 5, we provide a step-by-step approach to evaluating your course content for TALEs. Finally, the balance of the book looks at different class types and how our framework can be applied to each class to maximize the opportunity for a TALE, using examples and some helpful tidbits including active learning strategies. Remember, our goal is to not only create TALEs in study abroad and service-learning classes, but to create deep, impactful experiences in all types of classes – analytical classes, large lectures, studio and labs, and others. In our opinion, no class is inherently boring, inherently mundane, or too difficult for a deep experience; a class is limited only by the constraints of time and creativity. We hope that by structuring our framework in a step-by-step format, you can allow your creativity to run unbridled while we provide a mechanism to evaluate each activity.

So why give this book a shot? Because we, the authors – two faculty members and an administrator – know how hard this can be. Whether new to the field of higher education with a cursory understanding of teaching and learning fundamentals, or a seasoned veteran, we have attempted to situate this book to be beneficial to everyone. It is not a panacea, and it is not a substitute for planning and hard work. However, we have seen first-hand how passionate and dedicated AEC faculty and administrators are to provide high-quality learning experiences that prepare students to be leaders within their chosen fields. If you are one of these educators, we encourage you to keep reading and give our proposed approach a try. We hope your experience with this book is as thought-provoking and fun as it was for us to write.

2 The Significance of High-Impact Educational Practices

Do you recall a great learning experience in college that resonates with you like Sarah had in Chapter 1 – maybe it was an internship, group project, study abroad, unusual classroom experience? What was it? When did it occur? Why was that event impactful for you?

What Does Sarah's Experience Teach Us?

Sarah found incredible value in some very specific experiences during her time in college. In her class with Dr. Wake, the trip to Bolivia used a variety of pedagogical tactics that yielded educational, social, and personal memories. The trip was one week of her time in college, but the experience helped propel her toward a successful career and provided a framework for her professional life. Sarah firmly believes that her ability to communicate and problem-solve with a diverse group of people was introduced throughout her education but was solidified on her trip to Bolivia.

Sometimes as faculty, we hear something less than this outcome when people discuss their college educations. Students and families alike may even debate the value of the college experience. The challenge is certainly real; students, parents, and employers continue to demand more from colleges. And in this era of increasing educational costs and demand for qualified college graduates, universities need to consider what they can do to improve the student experience. More importantly, what can we (as educators) do to create those impactful experiences for our students that chart a trajectory for passion, success, and a desire for lifelong learning?

George D. Kuh (2008) set out to address some of these challenges and help students reach what he termed the "full benefits – economic, civic, and personal" of a college education. He recognized that universities were focusing on features like student retention, graduation rates, and time until graduation when, in fact, the real metrics important to impactful higher education may be different. Kuh's work has encouraged great discussion about what really matters in college education and has motivated deeper discussions about student learning outcomes and the

DOI: 10.1201/9781003106029-2

actual experiences that lead to those outcomes. Termed High-Impact Practices (HIPs), Kuh identified a specific set of practices that tended to lead to meaningful experiences for students.

How Did HIPs Develop?

The concept of HIPs was formulated as part of a larger decade-long program called Liberal Education and America's Promise. Developed by the Association of American Colleges and Universities, the program focused on clarifying the needs of employers in the 21st century. The goal of the program sought to create a higher return on investment for college students and their families while being inclusive and assisting students from diverse backgrounds to reap the full benefits – economic, civic, and personal – of the higher education experience. These goals were supported on a foundation of four essential learning outcomes:

- Knowledge of human culture and the physical and natural world (including study in the liberal arts)
- Intellectual and practical skills (including critical thinking, communication, teamwork, and problem-solving)
- Personal and social responsibility (including ethical reasoning, intercultural competence, and development as a lifelong learner)
- Integrative and applied learning

From these four outcomes, the Association of American Colleges and Universities set out to determine the greatest opportunities for achievement. Numerous research studies identified a core of experiences that provided substantial educational benefits to a diverse population of learners:

- First-year seminars and experiences
- Common intellectual experiences
- Learning communities
- Writing-intensive courses
- Collaborative assignments and projects
- Undergraduate research
- Diversity/global learning
- Study abroad
- Service learning
- Internships
- Capstone courses and projects

Researchers found that students who participated in such activities had higher grades and retained information at a rate higher than those who had traditional educational experiences. This was especially true for students who started out at a lower achievement level. HIPs provide an opportunity for students to connect information with specific experiences; blurring the boundaries between a student's academic life and out-of-class life and amplifying the social and cognitive impact of the experience (Terenzini, Pascarella, & Blimling, 1996). Students have an opportunity to reflect on their experiences, and confidence is increased as they return to the classroom or engage in professional practice.

Kuh introduced HIPs using specific teaching and learning practices. While delivery styles vary, almost all the practices involved active learning – an approach to teaching that continues to provide positive experiences for most learners. Kuh's work was built on educational research that showed an increased rate of student retention and engagement for each of the identified HIPs.

Since the introduction of the term HIPs approximately 15 years ago, HIPs have gained widespread acceptance across higher education (Kuh, O'Donnell, & Schneider, 2017). Entire state systems like California and New York have implemented initiatives to encourage students to participate in HIP-like activities (https://www.suny.edu/applied-learning/ and https://www.csustan.edu/hips). Additional research has shown that students who participate in multiple HIPs experience a cumulative effect, resulting in enhanced learning, personal development, and increased graduation rates (Finley & McNair, 2013). In short, HIPs are a regular part of the higher education vernacular and have demonstrated improved student success.

What Are the Limitations of HIPs?

Despite the identified benefits, Kuh's HIPs are not without controversy. In fact, Kuh himself acknowledges that offering just two of these experiences for each college student over their entire college career would be a major challenge. Criticisms are related to scope, value, and cost. First, the activities that could be used are limited to a certain type of class or format as identified in the actual "practice." Activities such as learning communities and undergraduate research are common but do not occur frequently over the course of a college degree. Thus, the scope limitation excludes a big portion of the curriculum included in most majors, or arguably worse, excludes a big portion of students, as many of the experiences require students to self-enroll.

Secondly, just because an activity is labeled as "high impact" does not mean that it is inherently impactful. Participating in an internship does not guarantee a deep learning experience. Certainly, an internship would qualify as a HIP, but if the student was assigned to answer the phone, sort files, or organize drawings, their experience lacks all forms of deep learning. And if they were never encouraged to reflect on their experiences, never connected with a mentor, and lacked basic exposure to real-world scenarios and problems, much of the experience is inconsequential. The student would not have truly "connected" and would have stopped short of a high-impact experience. Perhaps an internship, traveling abroad, or enrolling in undergraduate research does create a great opportunity, but opportunity does not assure success. For a high-impact practice to ultimately reap benefits, it must be constructed while considering deeper questions about student learning, equity gaps, and quality of delivery.

Finally, faculty can be reluctant to attempt HIPs because they can be onerous, requiring substantial resources. For example, Forbes reported that the cost of a study abroad program was double the cost of attending a private college for a semester when tuition was included (Davis, 2014). Such costs may deter students and families alike from participating in some HIPs.

Elements of HIPs

Despite the critics, HIPs continue to be academic drivers at a variety of higher education institutions. Perhaps, if we understood more about the details that led to Kuh's proclamation on HIPs, additional opportunity for use of the basic principles may exist. And, if we could employ those principles in a variety of learning situations, it may allow us to overcome some of the limitations noted above.

Kuh and his colleagues recognized eight elements that were common to HIPs. These elements included the following (Kuh, 2008):

Performance Expectations Set at Appropriately High Levels

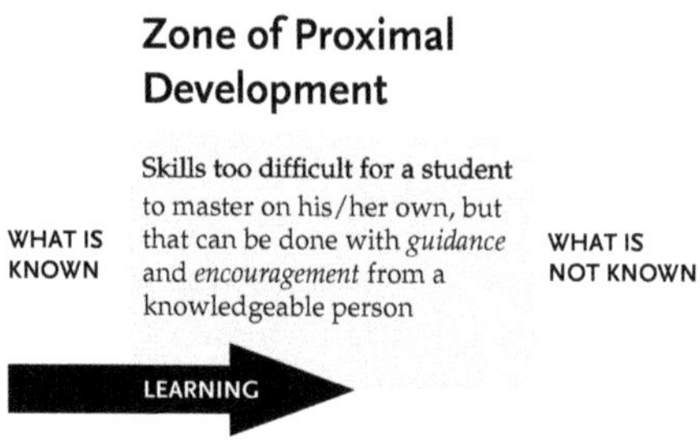

Figure 2.1 Appropriate Expectations Set the Stage for Learning.

By setting appropriately high expectations, teachers build a positive learning culture in the classroom that encourages students to see themselves as capable learners taking personal responsibility for their own learning. This idea follows the Zone of Proximal Development (ZPD) theory created by Lev Vygotsky (1978; Figure 2.1). When setting the level of expectation, faculty should pay particular attention to the knowledge, skill, and motivation the learners bring to the experience. A one-step-at-a-time level of improvement will allow students to progressively enhance their abilities instead of asking them to take giant leaps. As the learning process develops, some students may move outside their comfort levels. In these cases, faculty should offer new techniques or approaches that may bolster student success. While it is important to be attentive to underperforming students, care should be taken not to design activities to that level. Instead, instructors should focus on approaches, skills, and supplemental resources that provide opportunities for underperforming students to elevate their abilities.

Example: *Consider a design assignment where students are asked to place themselves in the eyes of the user of the facility and develop multiple narratives and sketches. This assignment would challenge students to achieve beyond current levels as determined by criteria set by faculty with input from the user of the facility.*

Significant Investment of Time and Effort by Students Over an Extended Period of Time

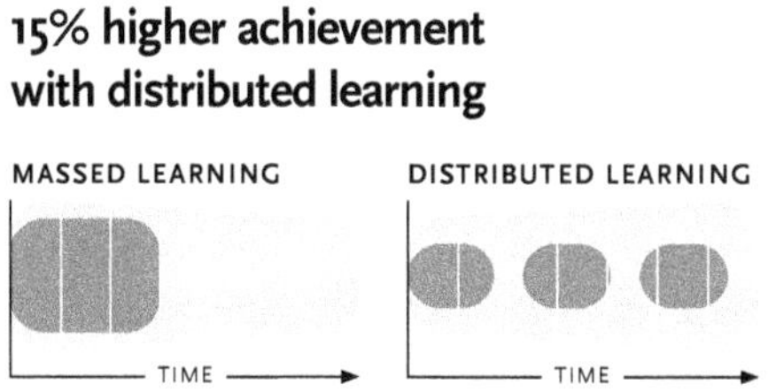

Figure 2.2 Achievement in Massed vs. Distributed Learning.

Known as "distributed practice," "spaced learning," or more recently, "scaffolding," the concept is simple. Multiple, spaced interactions with new knowledge, theory, or concepts lead to high-impact learning (Figure 2.2). Such an approach includes introducing a concept while providing multiple opportunities to reinforce the concept. Often, the best approaches use different activities to allow learners to interact with new concepts in a variety of ways. In his book, *Memory: A Contribution to Experimental Psychology*, German psychologist Hermann Ebbinghaus (1913) established the idea of the "spacing effect." Widely employed in education, the basic idea is that people are better able to remember information if it is learned in multiple sessions that are spread out over time. By simply employing spaced repetition, a person's ability to learn is enhanced.

Example: *Consider a multi-part class project where groups of students craft a response to a request for proposal for design and construction work over the course of a quarter or semester. The request for proposal may include a description of the responding company, descriptive narratives on design and construction, sketches of the design, site logistics plans, and cost/schedule of the work. Spread the deliverables out over the duration of the course rather than assigning all of them during the last two weeks of the course.*

Interactions with Faculty and Peers about Substantive Matters

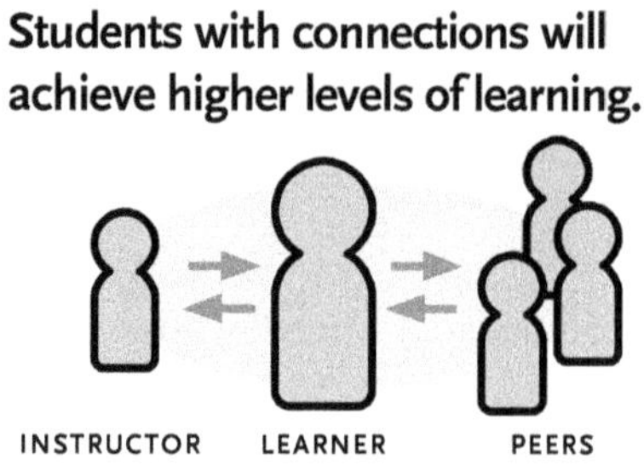

Figure 2.3 Interactions between Peers and Faculty.

Whenever an academic mentions the importance of interactions with others, Maslow's (1943) "Hierarchy of Needs" must be close at hand. Maslow identified that people need to feel a connection with others to address needs of achievement. Learning, and specifically, classroom learning, requires comfortable relationships with others in the room. Vygotsky (1978) explained how a student cannot reach a pinnacle of learning without the assistance of others. By placing students in situations where they interact with peers and faculty who are academically focused and goal-oriented, students are likely to mirror these features themselves (Figure 2.3). It is critical that these interactions be "educationally purposeful," tangibly seeking understanding and clarity. Learners are essentially steered to higher levels of learning through interactions with peers and faculty. This engagement leads to deeper learning and higher retention.

Example: *Consider a design history class that might discuss specific projects at least once per week and include input from faculty and students. Outside speakers may attend, and common readings may be discussed. Students may work collaboratively with faculty to create an assignment that developed from the discussion.*

Experiences with diversity, wherein students are exposed to and must contend with people and circumstances that differ from those with which students are familiar.

Figure 2.4 Experiences with Diverse People.

This HIP element implies diversity has at least two types. The first type refers to diversity of people. This structural diversity suggests that students should work with others from different racial or ethnic groups. The second type of diversity relates to the interactions students have during their time as a student. These interactions may be with ideas, with information, or with diverse people (Figure 2.4). Luckily for faculty, students often find their exposure to a diversity of information and ideas develops through connections they make with diverse people. Research shows that diversity enhances individual student development in the cognitive, affective, and interpersonal domains. In addition, recent research findings indicate that high levels of engagement with diversity in college lead to engagement with diversity after college. One of the tremendous benefits of working with others different than yourself is the tendency to think deeper about an issue. An article from Scientific American detailed, "Decades of research by organizational scientists, psychologists, sociologists, economists and demographers show that socially diverse groups (that is, those with a diversity of race, ethnicity, gender and sexual orientation) are more innovative than homogeneous groups." (Phillips, 2014) Interestingly, just the appearance of diversity (though different opinions exist) makes us think differently about how we approach issues. In short, diversity encourages students to consider different perspectives and reach stronger conclusions.

Example: *A service-learning project where students work with people in the community outside of what they would experience in the traditional higher education environment. The work would include reflection of the engagement by the students or perhaps a discussion session regarding the service-learning work and specifically the associated interaction with others.*

Frequent, Timely, and Constructive Feedback

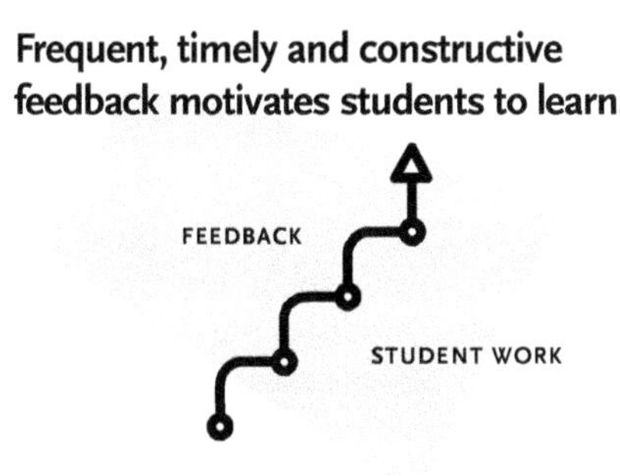

Figure 2.5 Feedback Opportunities.

"Feedback is a gift. It's data I didn't have before with which I can now make more informed choices" (Robin, 2013). Feedback is detailed information on the results of one's work. Formative feedback, which occurs over the entire class, is especially critical as students seek to understand how they can improve their ongoing effort. Instructors who provide prompt feedback followed by discussion of incorrect answers are using one of the greatest predictors of positive student outcomes. Constructive feedback is specific and should include both areas for improvement (corrective) and areas where the student has applied knowledge correctly (affirming) (Figure 2.5). Specific feedback allows clear understanding on how students can improve, while vague feedback often frustrates students leaving them unmotivated to improve their work. Positive encouragement drives students to continue to work hard and apply the knowledge they are learning.

Example: *In a structures class, consider assigning a problem after every new topic that would be due the next class period. The work would be graded either during the next class or immediately afterward with specific comments on errors or omissions made in the work. This feedback would propel the student to the next topic with one success building on another.*

Periodic, Structured Opportunities to Reflect and Integrate Learning

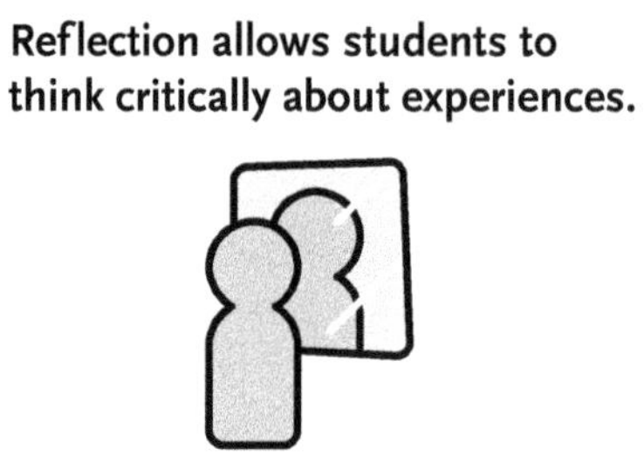

Figure 2.6 Reflecting on Learning.

Sports is a great "reflection" example. Football teams watch game film replay in the days following the game to process their actions in specific situations at a slower, more meticulous pace. The reflective period is important to student growth, and it allows for critical feedback in what seems like "real time" (Figure 2.6). Reflection is a key component that transforms many of the education experiences (service learning, study abroad, internships, collaborative projects, research) into learning. It can happen in a variety of ways including student presentations, writing, illustrating, speaking with others, and listening. In addition, the "structured" opportunities to "integrate learning" offer students the ability to rework assignments as a means of reflection and integration. As students rework what they've done wrong and perhaps provide a summary of their changes, deeper learning occurs as they attempt to better understand the material.

Example: *After a class that discussed how the speed of a vessel is proportional to the length of a vessel, have the students author a one-minute reflection paper in class on why the speed at which a 6'–6" tall swimmer might differ from a swimmer that was only 5'–10." Then, ask select students to share their insight.*

Opportunities to Discover Relevance of Learning through Real-World Applications

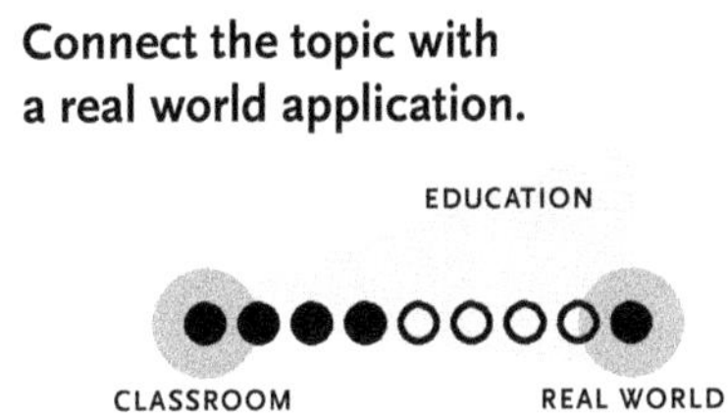

Figure 2.7 Real-World Relevance.

If you want to motivate someone, provide an opportunity to connect the topic with a real-world application (Figure 2.7). Students value the connection of what they are learning to their long-term goals. Learning that does not connect students with real-world applications has the potential to cause students to tune out and/or develop poor attitudes toward learning and school in general (Blumenfeld et al., 1991). When students have a clear answer to "why should I care about this" or "how is this relevant to my future career" their likelihood of interest and engagement increases. Key to this ability to be impactful for the employer is a student's experience with real-world, experiential learning. This type of learning allows students to develop skills while also relating to experiences outside the traditional academic setting. Interdisciplinary skills are promoted, leadership is fostered, and communication skills are strengthened in such an approach. All are highly valued by employers.

Example: *Consider an internship where students apply knowledge and skills obtained during their academic experience. During the internship, students should journal about the connection between their academic experience and their work in the internship.*

Public Demonstration of Competence

Education is too much like the sports equivalent of "practice." We need more "game day" experiences. A public demonstration of competence is exactly that – a game day experience. A student is asked to take what they have learned and convey that in an organized, logical manner to some interested group (Figure 2.8). Such a "demonstration of competence" can be defined as a practical event developed to evaluate an individual's (or a group's) ability to integrate and apply some combination of knowledge, skills, abilities, and attitudes necessary to be effective within the subject matter or area of practice. When we know that evaluation will occur and that multiple people will contribute, it often drives effort and expectation to present our best work.

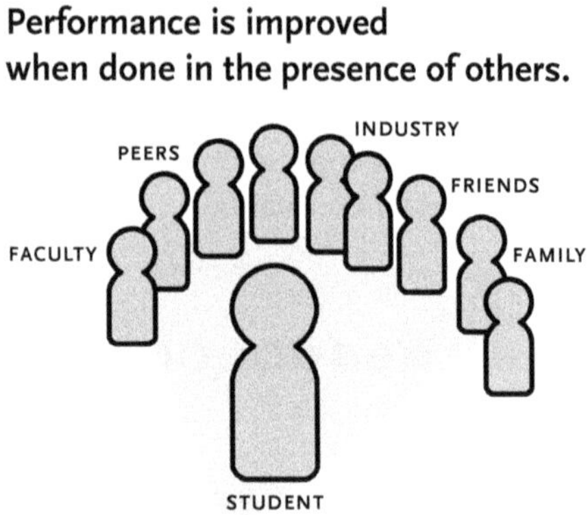

Figure 2.8 Public Demonstration of Competence.

Example: *After a graduating capstone class, students would make a formal presentation to classmates, faculty, and industry detailing the work conducted over the course of the semester. Feedback based on a predefined rubric may be provided with multiple perspectives of the audience.*

Is There Something beyond HIPs?

Since the publishing of Kuh's work in 2008, multiple efforts among faculty and researchers have been initiated to elevate the learning experience, and several universities are actively evaluating the implementation of HIPs and related impacts on their campuses. These efforts reinforce the recognition in higher education that there are opportunities and a need to improve the student learning experience.

We seek to use HIPs as an integral piece of a greater goal. We strive to create a repeatable process that will assure not only improved cognition and retention but also an impactful educational experience in a variety of settings beyond just those identified by Kuh. In our TALEs framework, HIPs have a roll, but instead of focusing on specific practices, this text uses the eight HIP elements mentioned above that yield a high-impact learning experience as part of a larger process, grounded in active learning. Our goal is to provide TALEs free of scope, value, and cost limitations.

References

Blumenfeld, P.C., Soloway, E., Marx, R.W., Krajccik, J.S., Guzdial, M., & Palincsar, A. (1991). Motivating project-based learning: Sustaining the doing, supporting the learning. *Educational Psychologist*, 26(3–4), 369–398. https://doi.org/10.1080/00461520.1991.9653139

Davis, A. (2014, July 10). 6 ways to cut the costs of your study abroad program. *Forbes*. https://www.forbes.com/sites/alexadavis/2014/07/10/6-ways-to-cut-the-costs-of-your-study-abroad-program/?sh=f8947d8ec326

Ebbinghaus, H. (1913). *Memory: A contribution to experimental psychology*. New York: Teachers college, Columbia University.

Finley, A.P., & McNair, T. (2013). *Assessing underserved students' engagement in high-impact practices. With an assessing equity in high-impact practices toolkit*. Washington, DC: Association of American Colleges and Universities.

Kuh, G.D. (2008). *High-impact educational practices: What they are, who has access to them, and why they matter*. Association of American Colleges and Universities. http://www.neasc.org/downloads/aacu_high_impact_2008_final.pdf

Kuh, G., O'Donnell, K., & Schneider, C.G. (2017). HIPs at ten. *Change: The Magazine of Higher Learning*, 49(5), 8–16. https://doi.org/10.1080/00091383.2017.1366805

Maslow, A.H. (1943). A theory of human motivation. *Psychological Review*, *50*(4), 370–396. https://doi.org/10.1037/h0054346

Phillips, K. (2014, October 1). How diversity makes us smarter being around people who are different from us makes us more creative, more diligent and harder-working. *Scientific America*. https://www.scientificamerican.com/article/how-diversity-makes-us-smarter/

Robin, C. (2013, November 27). *Feedback is a Gift* [Stanford University Graduate School of Business]. https://www.gsb.stanford.edu/insights/carole-robin-feedback-gift

Terenzini, P.T., Pascarella, E.T., & Blimling, G.S. (1996). Students' out-of-class experiences and their Influence on learning and cognitive development: A literature review. *Journal of College Student Development*, *37*(2), 149–162.

Vygotsky, L. (1978). Interaction between learning and development. In *Mind and society* (pp. 79–91). Cambridge, MA: Harvard University Press.

3 Moments Make a Difference

In a few sentences, describe a transformative learning experience while a student, that helped shape the person you are today (an experience that was both memorable and meaningful). Think about that event and the experiences that were a piece of it and note that the event could take place anywhere (classroom, internship, study abroad, work, athletics, fraternity/sorority, student government, etc.).

"The Power of Moments"

At one point or another, we've all had at least one experience in our lives that changed our outlook. Chip and Dan Heath's (2017) book "The Power of Moments" explores why specific experiences can have such extraordinary impact. Chip, a professor at Stanford's Graduate School of Business, and Dan, a senior fellow at Duke University's CASE center, have coauthored four *New York Times* bestsellers including *Decisive*, *Switch*, *Made to Stick*, and this book on moments. Their research shows that when we assess our experiences, we tend to recall flagship memories. These "Moments" are both meaningful and impactful. Sometimes, Moments happen informally. But others, for example, a high school graduation or wedding ceremony, are deliberate and require purposeful planning and effort over a significant amount of time. Moments like those are planned with the intention of creating a lasting memory for those who participate. The Heath's define Moments as, "meaningful experiences that stand out in our memory." In a sense, Moments create an emotional attachment to an event or period of time. While HIPs focus on deep learning for students in an academic setting, Moments can occur at any time, in any place. Using the Heath's literature as a guide, we can take the constituent parts of a Moment – the "components" – and apply that information to an academic setting. In future chapters, we will show that by combining the components of Moments and the elements of HIPs with other strategies, we can produce deep, emotional, and active experiences.

Moments matter intensely to students. Do you remember what it was like when you were selecting the college you were going to attend? If you're like us, you were filled with excitement at the makeup of the campus, the sporting events you would attend, the dorm arrangements, friendships you would make, parties you would attend, and all the other social opportunities synonymous with the

DOI: 10.1201/9781003106029-3

college experience. Students go to college in expectation of experiences they will retain for a lifetime. Today, many students spend much of their time working to reduce student loan debt, attending university-sponsored events, participating in extracurricular activities, and engaging in social programs important to them. This type of involvement is fantastic for the student but can leave less time for impactful academic pursuits. While the standard has always been to spend about three hours outside of class each week for every credit hour taken, the average college student (across all disciplines) today spends about one hour. Considering most students take 12–15 credit hours of coursework per semester, this means less than 20% of their time is spent on schoolwork outside of class. This means time in class has become that much more important.

You're probably asking yourself, "Do these jokers really think it's possible to get students as interested in their classes as a Friday night party, an intramural sports team, or the big football game this weekend?" No, we are not proposing students will prefer academics over these other activities if we make our classes more interesting. We are proposing with some thoughtfulness and planning, we could create opportunities that elevate the learning experience to be as equitable for the students as these other college experiences – learning experiences that create memories that will last a lifetime – Moments. The Moment we describe is a point at which many educational practices and experiences converge to connect the learner to a realization that something meaningful and personally relevant has happened.

Another Way to Think about Moments

One of the key items the Heath brothers note is that school and our educational system need to be more like sports, and in particular, game day. Sports are social, exciting, and have a heightened level of engagement that is difficult to replicate in everyday tasks. Instead, school often mimics "practice," without the prospect of a game day. Practice is drilling and repetition, but without a game to look forward to, practice is grueling and mundane. Perhaps basketball great Alan Iverson captured the student perspective on practice best in a now famous interview quote when asked about his effort level during a practice:

> … I mean, listen, we're talking about practice, not a game, not a game, not a game, we talking about practice… Not the game that I go out there and die for and play every game like it's my last. Not the game, but we're talking about practice, man. I mean, how silly is that?

Perhaps the educational experience could improve if we, as educators, structured the learning experience to have a few less practices and a few more games. As humans, we naturally elevate our performance more when it's game time. If the goal is to create transformational active learning experiences, we need more game days.

What Conditions Are Necessary to Create a Moment?

The Heath's research concluded that when a person has an extended experience like college or even a 15-week class, they tend to remember only a few key aspects from that experience. Psychologists note that people tend to remember (1) the best or worst event and (2) the ending. Beginnings, endings, and times of transition are particularly ripe for teachers to create meaningful educational Moments, or to "build peaks" as the Heaths say.

These flagship experiences must be intentional and purposefully designed by faculty. Consider, for example, a class that has a culminating presentation on a significant project. The default approach has the students present to the faculty member. A more impactful approach has the students present to their peers in class. The most impactful approach is to invite an industry panel who had personal knowledge or interest in the problem being presented. The decision to integrate an industry panel significantly enhances the experience not only for the students but also for the faculty and panel members.

In another example, consider a structures class that requires students to determine how much weight a small concrete beam can hold before failing. The students will certainly run calculations in order to understand basic beam design, but by adding a hands-on lab, with either scale or full-size concrete members, students can witness the result of a beam that fails at mid-span in tension. Moving the experience away from being strictly mathematical will almost certainly give the students a heightened appreciation for why rebar is important to a concrete beam. These examples illustrate intentional design of peaks in the student learning experience.

The Heath's research determined that Moments were composed of one or more of four key components – Elevation, Insight, Pride, and Connection. How could we explore and collate a series of components that would create a Moment? Imagine comparing this to climbing a mountain (Figure 3.1). In the beginning, this will seem like a challenging task that is certainly different from everyday life, and it will take time to move toward the top. But, as you experience traversing the peaks and valleys of the mountain you begin to learn new things about yourself, build pride in what you're accomplishing, and develop new connections with those who have experienced something similar. This process can lead to a meaningful educational Moment.

Elevation

The concept of "elevation" is an experience that is perceived as something greater than everyday life – in a sense, exceeding the individual's expectations. Consider the examples of what incorporating the four subcomponents of elevation may include:

- *Boost sensory appeal* – A field visit to view a mock-up panel replaces a lecture on waterproofing.

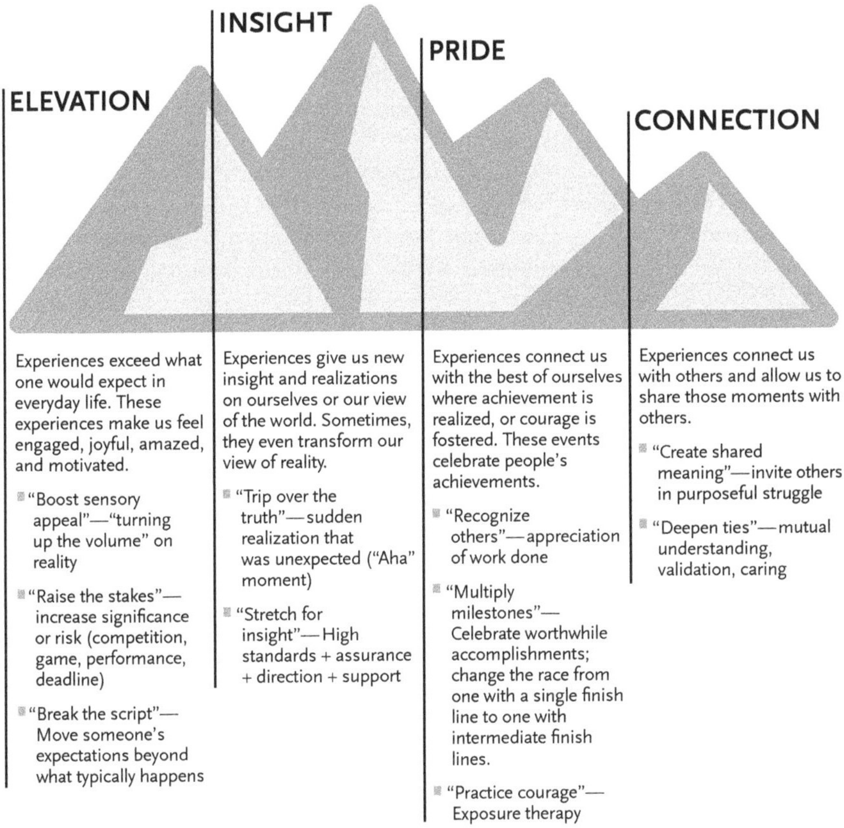

Figure 3.1 Components of a "Moment"

- *Raising the stakes* – A typical classroom assignment is converted into a competition between student groups.
- *Breaking the script* – A lecture on career opportunities is replaced by short discussions between people who hold those jobs and students.

Insight

Have you ever had an experience that caused you to rethink your perspective of something? This is what "insight" is all about. The Heath brothers suggest there are two ways to create insight – *tripping over the truth* and *stretching for insight.*

Tripping over the truth could be compared to the analogy, "You can't see the forest for the trees." Sometimes we have become so close to something that we are unable to see the bigger picture. The McWhorter School of Building Science at Auburn University has a capstone class that requires students to apply all the knowledge they have acquired to a construction project. Among other things, they develop an estimate, create a safety plan, develop contracts and cash flow documents, complete a structural analysis and schedules, and create a fictitious company to complete the work. Almost everything they are asked to do forces them to "create" from skills and knowledge developed over their four-year curriculum. Students generally find that their specific example is different than what they learned in class, and they are forced to reflect on their skills and apply them in a slightly different context. Students find they trip over the truth that they do have the ability to apply the skills they have learned throughout their time on campus. It's a process that has proved invaluable for thousands of graduates, and one that many graduates would argue has defined the construction management program.

Stretching for insight is a willingness to take risks for the betterment of expanding your self-understanding. Consider the story of Kia who was excited to enter her first studio. She had always loved design but shied away from math and science classes. To her dismay, the first major project in the studio seemed to be numbers-based. Dimensions, proportions, and a requirement to be "exact" caused her stress, and she wondered if she was in the right place. Her lead faculty member, Ms. Speers, set high expectations that the skills taught be implemented exactly. And she buoyed those expectations with encouragement and assurances. "Kia, I expect you to get this correct, and I know you can," was a phrase uttered multiple times over the studio's first few weeks. When Kia failed, Ms. Speers often repeated the exercise or a smaller version of it with Kia providing clear direction and support. Ms. Speers knew that the work requested was new and unnatural to the students including Kia, and this required each of the students to take small risks. This risk, in turn, would provide insight. As the first studio wrapped up, Kia looked back knowing she had grown through the experience, grateful she had persevered through the process.

Pride

Educationally, there are many things an instructor can do to instill pride. Examples vary from setting high standards, displaying work publicly, recognizing accomplishments, to completing a challenging project or assignment. These are the situations that cause students to stand a little straighter and hold their heads high. Consider the examples of what incorporating the four subcomponents of pride may look like:

- *Recognize others* – At the culmination of a semester-long project, incorporate a recognition ceremony.
- *Multiply milestones* – Break a larger class project into smaller, structured deliverables students submit as they work toward completing the project.
- *Practice courage* – Encourage students to make low-stakes presentations in class ahead of a larger, more formal presentation.

Connection

At this point it's quite easy to see how Elevation, Insight, and Pride can create resonating Moments, and we've taken extra effort to show how they can be impactful even on their own. You've likely experienced some of these in your life. Perhaps some of your most impactful Moments were with others. This is what Connection is all about. They are those experiences that allow us to develop a *shared meaning* with others that cause us to *deepen ties* with those people.

Many architecture programs teach using a studio environment. Students are challenged to develop an individual solution to a single problem that the collective studio seeks to resolve. Students frequently ask each other questions and have critique days where faculty inquire about approach and design. Over time and repetition of multiple concepts, *deep ties* are often built between students and the faculty evaluating them. Students value others that are engaged in the daily grind of the studio (*shared meaning*) and have a mutual understanding and caring for the work. The studio is an inherently social environment, and it is more memorable because students experience it together. Well beyond graduation, architecture students will often recall their studio experiences.

Connection is not just about putting people together to accomplish a task. It requires a connection so that students work toward a larger sense of meaning – toward a purpose. Perhaps they are solving a problem for a specific person or group of people; perhaps they are addressing a real-world problem faced by the industry; regardless, this solidifies the impact of the connection.

Moving beyond Moments

Perhaps you are having a tough time with this concept. We get it – tying education to emotion is a complex topic. What happened to the noble pursuit and acquisition of knowledge? That's precisely the issue, as educators we've already been impacted by profound Moments that have developed passion that fuels our desire to learn and educate. Unfortunately, we often expect our students to show up with the same passion, even though they likely haven't experienced any profound, passion-instilling educational Moments. It is easy to forget what drove us all to become educators. The excitement we feel for our discipline was almost certainly fostered by others, perhaps even by one or more of our college professors. If you are still having trouble with this idea, there is a practical side to all of this – passion fuels motivation to learn.

Ultimately, Moments matter to students. What students need are a few more games and a few less practices. We propose that with intentional design for Moments, active learning techniques, and the elements of HIPs, the perfect environment for a TALE can be crafted.

Using Moments

To summarize what we have learned to this point, there are three conditions necessary to create Moments for students. First, timing must be considered. Beginnings, endings, and periods of transition in learning are the highest opportunities for a Moment. Second, teaching well is difficult, and creating Moments requires purposeful effort and a consideration by the faculty to project where the "peak" experience may occur in the learning process. Finally, employing one or more of the four key components identified by the Heaths – elevation, insight, pride, and connection – provides a foundation on which to build a Moment. While it is possible to have a defining Moment with only one key component, the more that can be achieved, the greater the chance for a Moment.

There are appropriate times for Moments to happen. We need to be strategic in creating these Moments for our students so that they do elicit the impactful outcome we are intending. If you think about your class as a guided tour through an educational experience, as the creator of that experience, you are the one responsible for establishing the route. If you want that tour to be impactful and memorable, you must think about what that journey looks like through their eyes.

As educators, we recognize there is no specific "Moment" of education. Education begins when we are born and continues until we die. No single college experience (test, paper, or project) serves as one's education. Education in a college environment is a process that works only with appropriate time and the right kind of effort. Reducing the idea of education to a single Moment is discounting the tremendous value our universities contribute to the learning process. We should be clear; all learning does not happen at the "peak Moment," as we gaze off at the horizon we have mastered. Much of the learning process requires hard work by students and faculty alike. Often this setting of expectations may allow one to see the mountain top in the distance but requires substantial time traversing the mountain before the peak can be reached. But the mountain peak is an opportunity for a grand Moment.

Reference

Heath, C., & Heath, D. (2019). The power of moments: Why certain experiences have extraordinary impact. Corgi.

4 A Foundation for Transformational Active Learning Experiences (TALEs)

What would the intersection of HIPs and Moments look like? How might it make for a more robust learning opportunity?

Clarifying HIPS and Moments

In order to fully understand transformational active learning experiences (TALEs), we want to be sure that the differentiation between high-impact educational practices (HIPS) and Moments is clear. While working to produce your TALE, it is helpful to think of these two items as separate but equally powerful tools. We are not suggesting that HIPs and Moments do not have any overlap, but as we begin to build this concept of a TALE, it is helpful to think of them separately. Using the beneficial pieces within each of them provides a solid foundation for a TALE.

HIPs are – as the name states – practices – strategic, programmatic learning exercises meant to foster deep learning. These are intentional in design and are based on educational research. Moments relate to the emotion we attach to events or periods of time. Moments lead to lasting memories. For an event or experience to formulate into a lasting memory, there is likely a heightened sense of cognition about what is occurring during the event or in reflection after the event. This enhanced cognition leads to retention. The benefit to a student experiencing heightened cognition and retention is obvious and what we intend to harness.

We want students to experience deep, emotional, learning in all class types – even the ones that would historically be deemed "boring." This is our goal, and TALEs is our approach.

Limitations in Both HIPs and Moments

In Chapters 2 and 3, we noted several benefits to employing HIPs and Moments. For all the good that comes with these, we also noted limitations. For HIPs we outlined key limitations:

- HIPs are limited to a few, specific activities (practices) that don't occur every semester for the average college student.

DOI: 10.1201/9781003106029-4

- HIPs tend to be focused on a few specific class types and tend to be self-selective by students.
- HIPs aren't inherently "high impact" just because the experience is labeled as such.
- HIPs can be onerous for faculty, requiring time, money, people, space, etc.
- HIPs can require significant cost to students.

For Moments, we noted that:

- Moments are not specifically for education and therefore require extrapolation.
- When Moments are experienced, often, the preparation that went into the event or experience to create the moment is not recognized.
- Creating true Moments requires attention to multiple components of elevation, pride, insight, and connection.

Taking the Good, Leaving the Bad

What if we could use components of Moments to address some of the limitations of HIPs, and use the elements of HIPs to assure that the Moments we create are impactful learning experiences? In such an approach, we could create a variety of experiences – grounded in active learning – that could happen in almost any type of class, at any point along the college journey. Combining what we know about Moments, and using the elements of HIPs, we are setting the stage for TALEs. We will talk a lot more about how to create TALEs in the next chapter, but for now, let's spend time analyzing known TALEs.

The simplest way to analyze an achieved or potential TALE is graphically (Figure 4.1). The horizontal columns indicate the eight identified elements of HIPs. The vertical components suggest how Moments are created. By creating intersections of one of the four components of Moments and one of the eight elements of HIPs, we develop a purposeful foundation for a TALE. By graphically mapping the intersections of HIP elements and components of Moments, we now have a metric for measuring the likelihood of a TALE prior to the activity and a mechanism for reflection once the activity is executed. As the authors, we believe that the most powerful TALEs have some combination of HIP elements that touch all four components of Moments.

TALE Strength Summary

Where do the HIPs and Moments intersect in your proposed plan?

Elements of HIPs	Components of MOMENTS: ELEVATION	INSIGHT	PRIDE	CONNECTION
What will you do to **set the expectations** so they are challenging for the students but not discouraging to the point they give up?				
How will the experience be set up to require a **significant investment of time and effort** by the students?				
How can the experience involve the students having **interaction with faculty and peers**?				
What can be done to have students **interact with others** that they wouldn't normally interact with in the traditional class experience?				
What aspects of the experience will **provide** students with **frequent, timely, and constructive feedback**?				
How can you incorporate **periodic opportunities within the experience for students to reflect** and integrate what they've learned?				
How can you make the experience feel more **real-world** for the students instead of a ficticious learning exercise?				
What can you do to have the **students demonstrate their knowledge and skills** to others outside the class (public display)?				

X *Indicates an area where HIP element and Moment align*

Figure 4.1 Intersection of HIP Elements and Components of Moments.

Revisiting Sarah

Consider Sarah from Chapter 1. Let's revisit her experience in the context of what we're proposing:

> The Bolivia trip was only a week, but what a week it was. They had spent much of the semester designing a water irrigation system for a remote village, an 8-hour drive from their initial destination, La Paz. Leading up to the trip, the group of ten students had really grown to trust in each other. Their professor, Dr. Wake, spent a great deal of time preparing them for the experience, and set clear expectations to create a workable irrigation system for this arid community. On arrival to the tiny village, Sarah was met by one of the local children who insisted she spend time with her coloring a picture. At a service to celebrate their arrival, Sarah heard several of the locals share (through a translator) what the project would mean to the community. As she prepared to start the week, it became very apparent the challenge that lay ahead. The altitude, the heat, and the available tools all seemed daunting to the task. However, she loved working side-by-side with the Bolivian locals and getting to learn more about each of them and their culture. Edwardo, who said very little but worked tirelessly, showed her some tricks to make the work easier. Over the course of the week, she could tell he was drawn to the students' passion for the work, and a connection was built that extended beyond languages. Before leaving, the villagers cooked a special dinner for the students, and each student was asked to share their experiences. Sarah vividly remembers the smell and taste of that meal, the passion in her voice as she shared her experience, the pride in completing the project, and the relationships she fostered with both students and villagers.

Dr. Wake had crafted intersections of HIP elements and components of Moments (Figure 4.2). When we graphically evaluate the example, we can shade in the areas of intersection spots where HIP elements are being achieved within the context of Moments.

Elevation and Pride

The elevated aspects of the trip occurred at the dinner which was both a Moment of elevation and a Moment of pride. The students could reflect and integrate their learning through discussions on the impact of the irrigation system and discussions about what they could have done differently during the week. The dinner also culminated a long week of application that had a real and profound impact on a community. For the students, the dinner "broke the script" of what was expected in a remote, mountaintop village in Brazil, and the event "boosted the sensory appeal" through a special meal prepared especially for the students. Pride was realized by the students by simply being "recognized" in such an unexpected way.

TALE Strength Summary

Where do the HIPs and Moments intersect in your proposed plan?

Elements of HIPs	Components of MOMENTS: ELEVATION	INSIGHT	PRIDE	CONNECTION
What will you do to **set the expectations** so they are challenging for the students but not discouraging to the point they give up?		X		
How will the experience be set up to require a **significant investment of time and effort** by the students?		X		
How can the experience involve the students having **interaction with faculty and peers**?				X
What can be done to have students **interact with others** that they wouldn't normally interact with in the traditional class experience?				X
What aspects of the experience will **provide** students with **frequent, timely, and constructive feedback**?				
How can you incorporate **periodic opportunities within the experience for students to reflect** and integrate what they've learned?	X		X	
How can you make the experience feel more **real-world** for the students instead of a ficticious learning exercise?	X		X	
What can you do to have the **students demonstrate their knowledge and skills** to others outside the class (public display)?				

X *Indicates an area where HIP element and Moment align*

Figure 4.2 Intersection of HIP Elements and Components of Moments for Sarah's Story.

Insight

Dr. Wake set high expectations for the group and engaged them through substantial time and effort over the course of the semester. The goal of designing and installing a water irrigation system would be daunting for any college student, but Dr. Wake's knowledge coupled with a trust in the students, created a Moment of insight. As work progressed, the students became more confident in their capabilities. Successes in small, sequential assignments propelled students toward the design of the full irrigation system. Dr. Wakes' appropriately high level of expectation and their action to complete the design and install the system had pushed his students to "stretch for insight."

Connection

The study abroad experience was carefully crafted to engage the students with the people of Bolivia. This experience with diversity led to a Moment of connection where students had a greater appreciation for the Bolivian villagers, and the villagers had a greater appreciation for the students. This cultural interchange required students to move outside the walls of the academic experience and engage with the world around them in a unique way.

Summary of Intersections

Sarah's example was what would be considered a pinnacle example of a TALE, built on a classic HIP. But, sending students to another country to achieve this type of experience is not an option for all students. The goal of TALEs is to provide an approach that creates an opportunity for all students – even in the simplest of settings. So, let's consider a more basic approach that seems more achievable by any teacher in any class on any campus.

TALEs in the Everyday

We believe that there are countless ways to deliver a TALE, and we encourage you to think about TALEs broadly. Certainly, a specific event (trip abroad or other milestone event) provides a more obvious opportunity for a TALE, but TALEs can occur as part of a typical class if an instructor can connect deep learning (Elements of the HIPs) and times of consequence (Moments). The connection of HIP elements and components of Moments in the creation of a TALE is often enhanced through active learning strategies. Throughout the book we will present TALEs using real-world examples. While reading, you will note that the creation of a TALE and the use of active learning are inextricably linked. Active learning enhances cognition and retention through participation. If students are not engaged in their learning, creating an emotional (Moment) and intellectual (HIP elements) connection to content becomes difficult. Several active learning strategies are identified in Figure 4.3 (Brame, 2016; *How Can You Incorporate Active Learning in Your Classroom?* n.d.).

Active Learning Strategies Defined

Clarification pauses/pause procedure: After presenting a key point or concept, pause and give the students time to think about the issue, then ask if anyone needs anything clarified. One way to do this is to ask students to review their notes during the pause.

Retrieval practice: Pause for an appropriate amount of time every 15 minutes, having students write or execute everything they can remember from the preceding class segment. Encourage questions.

Student self reflection/"One minute paper": Students reflect on what they've learned by authoring an approximately 1-minute response to a question posed by the teacher.

Self-assessment: Students complete a quiz (typically ungraded) to determine their competency level of the subject.

Large group discussion: In groups of 10 or more, students discuss questions presented by the teacher in regards to a reading, video, or topic introduced in class.

Think-pair-share: Teacher poses a question to the class; students formulate an individual response; and pair up with someone in their proximity to discuss their responses. Critical thinking is fostered. Often think-pair-share will lead to group discussion followed by whole class discussion.

Cooperative groups/Informal groups: Consider forming groups of 3-4 students to use throughout the semester. During a lecture, pose a question for the groups to discuss between themselves. While groups are discussing, the teacher will walk the room stopping at several groups to keep the groups focused on purpose. Teachers can ask groups to present their findings to the class.

Demonstrations: This approach asks students to predict an outcome for a demonstration that they will witness. Once the demonstration is completed, students will discuss what they observed, anything that was different from what they predicted, and suggest what may have caused the outcome to be different than what they predicted.

Peer review/Peer evaluation/Group evaluation: Similar to think-pair-share, peer review allows students to participate in a low-stakes review of work while engaging in higher-level cognitive skills by analyzing an evaluating the work of others.

Brainstorming: Students work in groups to quickly generate idea responses to a question, topic, or problem. This is a good technique to use at the beginning of a new topic when trying to assess student's current level of knowledge and understanding.

Case studies: Use detailed accounts of situations that occur to people, projects, families, events, etc., that prompt students to connect what they've learned with real-world situations, actions, and consequences.

Figure 4.3 Various Active Learning Strategies Used to Support TALEs.

Active Learning Strategies Defined (Cont'd)

Hands-on technology: Students use technology such as software, visualization, or communication devices to enhance understanding of concepts being studied.

Interactive learning: Find a time during the lecture to include an activity that allows students to work with the material being covered. This could include the students solving a problem posed, exploring graphics, or connecting the material to professional practice.

Strip sequence: Students are provided steps of an ordered process on sheets of paper in random order, then asked to construct the appropriate sequence.

Concept maps/Mini-maps: Students create connections between parts of a concept to visually illustrate their understanding. Some examples might include steps in a process or relationship of people in an organization or team.

Active review sessions (games or simulations): The instructor poses questions or deliverables, and students work on them in groups or individually. Responses are presented to the class for evaluation and discussion. This can be competition based.

Role playing: Students assume the role of a person or a position as part of an "act" that accurately demonstrates a situation and/or people's roles in those situations. Preparing for the "act" helps students enhance their understanding of the concepts and theories being discussed. Executing the "act" demonstrates their level of understanding. Feedback from the audience after the act provides immediate feedback to the performers.

Jigsaw discussions: A general topic is divided into smaller, interrelated pieces like a puzzle. Each member of a team is assigned to become an expert on a different piece of the topic. After each person has become an expert on their piece of the puzzle, they teach the other team members about that piece.

Inquiry learning: Students use an investigative process to make observations, pose hypotheses, and speculate on conclusions about questions posed by the instructor. Then students share their thoughts and tie the activity back to the main idea/concept.

Forum theater: Students are the audience and actors as in a theater. The situation can be played out multiple times with different students taking different roles and perspectives.

Experiential learning: Activities that allow students to observe and apply concepts discussed in class to a real-world situation.

Figure 4.3 (Continued).

In subsequent chapters, we will highlight select strategies from this list and how they are applicable to the class type being presented. These are nonexhaustive, but a good place to start.

Carlos' Experience

In the spring of 1988, Carlos took Physics II as a sophomore at Duke University. The faculty for the course was a young physics professor, Alec Joules. Dr. Joules connected well with the students and made himself available to them frequently. His love of the subject matter was obvious to the students. Somehow, he found ways to explain electromagnetism in at least a half-dozen ways. To Dr. Joules, teaching was important, and he believed that each individual student, regardless of experience, could learn.

Dr. Joules' class was not easy. The homework was always difficult and stretched over numerous class periods. Despite the complexity, the homework had a way of scaffolding knowledge together that built a foundation from the material, and the conclusion of an assignment always ended with in-depth discussions. Dr. Joules' initial crafting of the TALE was simply to give students opportunities to learn and find their own success through trial and error. Low stakes homework and lab exercises provided opportunities for students to make errors, analyze the issue, make a change, and try again. Much of this happened before a single grade of significance was recorded for the class.

Dr. Joules facilitated this scaffolding of homework through a variety of class practices. First, he was an expert at show and tell. Many of us in the Architecture, Engineering, and Construction (AEC) profession learn by seeing as opposed to hearing about it (modeling). For Dr. Joules, modeling was the cornerstone of the scaffolding process. Every chance Dr. Joules had, he demonstrated for students what they were to do. Often, he would solve problems on the board only to follow them with a "think aloud" where he illustrated or modeled his thought process as he solved problems. Second, he used visual aids to guide and shape students' thinking. Often, this visual aid would break the script moving away from traditional formulas and toward an application of concept. Finally, he used an impactful strategy while modeling problems where he paused, asked questions, paused again, and then reviewed the work completed. The pauses provided time for students to process information and think, and the questions were more strategic focusing on reflection of the learning. In one case, a student might be called on to give a summary of what was just discussed or outlined. And, if the group was stuck on a question, he would encourage them to chat with their neighbor to discuss the issue. Such an approach helped keep students in his physics class active and engaged participants.

Dr. Joules had effectively crafted a deep approach to learning. High expectations coupled with repeated conditioning of complex problem-solving over the course of the semester were a recipe that impacted students. Since problems built continuously on concepts learned during the semester and often involved real-world applications, students soon developed an internal passion to solve complex problems instead of just being motivated by the grade.

After finishing several homework assignments, Carlos started to believe that he could accomplish incredibly difficult tasks. The homework was purposeful, required inclusion of lessons that were acquired throughout the class, and always required the application of technical concepts to real, tangible problems one might find in engineering practice.

The culmination of the class led to an event that was a true TALE for Carlos – his final exam.

At Duke, final exam week was always a meaningful and definitive experience. Carlos remembers spending hours preparing for exams with his colleagues – study groups, index cards, discussions. This was especially true for Professor Joules' class. Knowing the final exam was the culminating piece of his TALE approach, Dr. Joules regularly referenced it throughout the semester. As the exam neared, Carlos and his fellow classmates were absolutely committed to displaying how much they had learned. They knew that the exam would force them to apply what they had learned in a different setting and stood as a seminal milestone in their educational physics experience. There were 30+ questions on the exam, but all these years later, only the last one stands out as Carlos thinks back to his undergraduate education:

> You are stopped by a highway patrolman for running a traffic light. Instead of just accepting the ticket for running the traffic light, you present your case to the officer that you were driving so fast that the Doppler shift made a red light appear green. The officer agrees to give you a pass on running the traffic light but tells you that you will receive a speeding ticket with a fine of $1 for each mile per hour you were exceeding the 45-mph posted speed limit. How much was the fine?

Connecting HIP Elements and Components of Moments in the TALE

Professor Joules' class could be visualized in Figure 4.4.

Elevation

Professor Joules wrote a final exam that students will always remember. While many of the questions on the exam tested basic understanding and application of knowledge common in most core science finals, Dr. Joules found a way to elevate the experience by one specific question geared toward a more comprehensive understanding of the topics within the class. Few, if any, undergraduate sophomores had ever seen something that required one to apply knowledge in such a direct way. The problem "boosted the sensory appeal" (Moment component) for the students through a discovery of relevance through a real-world application (HIP element).

Insight

The problem allowed each student to view themselves as someone who could solve ill-structured problems. Dr. Joules had set the expectation for high standards

TALE Strength Summary

Where do the HIPs and Moments intersect in your proposed plan?

Elements of HIPs	Components of MOMENTS: ELEVATION	INSIGHT	PRIDE	CONNECTION
What will you do to **set the expectations** so they are challenging for the students but not discouraging to the point they give up?	X	X	X	X
How will the experience be set up to require a **significant investment of time and effort** by the students?	X	X	X	X
How can the experience involve the students having **interaction with faculty and peers**?	X	X		X
What can be done to have students **interact with others** that they wouldn't normally interact with in the traditional class experience?	X		X	X
What aspects of the experience will **provide** students with **frequent, timely, and constructive feedback**?		X	X	
How can you incorporate **periodic opportunities within the experience for students to reflect** and integrate what they've learned?		X	X	
How can you make the experience feel more **real-world** for the students instead of a ficticious learning exercise?	X	X		
What can you do to have the **students demonstrate their knowledge and skills** to others outside the class (public display)?	X		X	

X *Indicates an area where HIP element and Moment align*

Figure 4.4 Intersection of HIP Elements and Components of Moments for Carlos' Story.

(HIP element) throughout the homework and the course. By solving those problems in small versions and being available to help, he allowed the students to build confidence while receiving direction and support (HIP element). By the time the final arrived, and a student faced this exam problem, the insight he helped the students develop provided them a background to solve this problem. The students experienced "self-discovery" (Moment component) that he or she could solve a tough, applied problem.

Pride

Many students walked out of that exam confident that they had provided a reasonable, thoughtful response based on the tools given in the class. Students knew they could tackle hard problems, and the next problem students faced would be made more solvable because of that experience. The confidence the students had to complete such a complex exam came from the structured engagement with Dr. Joules (HIP element) built in throughout the semester, addressing real-world problems (HIP element). Difficult homework, culminating in a difficult exam, gave the students the opportunity to practice courage throughout the semester (Moment component). Complex questions often require bold responses. Dr Joules instilled courage through a culture of encouragement and recognition (Moment component).

Connection

Throughout the semester, the students created a shared meaning (Moment) in the classroom by solving challenging homework problems together (HIP). The result was a class with mutual care for one another.

Summary of Intersections

Let's think about these intersections between elements of HIPs and components of Moments a little differently. In essence, we are using the elements of HIPs and active learning to build various moments of elevation, insight, pride, and connection. In this approach, specific notes and lectures matter less than how the instructor comprehends the subject and values deep, engaged student learning. This develops a purposeful foundation for the implementation of what could be a TALE. In the next chapter, we'll walk through applying the framework to build a TALE, and allow you to begin considering how your own TALEs may develop.

References

Brame, C. J. (2016). *Active learning.* Vanderbilt University. https://cft.vanderbilt.edu/wp-content/uploads/sites/59/Active-Learning.pdf

How can you incorporate active learning in your classroom? (n.d.). [University of Michigan Center for Research on Learning and Teaching]. Retrieved July 1, 2021, from https://crlt.umich.edu/sites/default/files/Active_Learning_Continuum_CRLT.pdf

5 Building a TALE

> Where could you imagine a TALE in your current class? What aspects of your current class do you see that could be enhanced or expanded?

In Chapters 2 and 3 we talked about elements of High-Impact Educational Practices (HIPs) and components of Moments in the context of strategies to heighten the student learning experience. In Chapter 4 we suggested that combining the two strategies would create the perfect storm of learning experience in what we called Transformational Active Learning Experiences (TALEs). This chapter describes the process for maximizing the opportunity for a TALE.

Before we jump into the approach, we want to clear up a couple of things in this chapter. First, it's worth noting that to successfully implement the TALE approach, background knowledge in fundamentals of curricular design (learning outcomes, assessments, and instructional development) is helpful. For those who are unfamiliar or need a refresher on these concepts, we have included a brief overview in the appendix to this chapter. Having a basic understanding will make implementing the strategy explained in this chapter much easier and increase your chances of success. Second, we will use the terms "activity" and "experience" interchangeably. When we say "experience" or "activity," we are alluding to the learning exercise that you are developing into a TALE.

TALE Framework

The conceptual framework for building a TALE is broken into six steps, illustrated in Figure 5.1. While this chapter is descriptive in how to execute the framework, the next chapter will apply these steps to an actual class example.

It should be noted that many of the steps are situated for faculty that have taught a course before and are adjusting an existing activity; however, if you are teaching a course for the first time, the approach still holds value. In this case, where the steps ask to evaluate the "current" approach, use this as your initial approach. Regardless of your experience with TALEs, we suggest starting small and growing into a well-developed TALE. Make incremental adjustments to the

DOI: 10.1201/9781003106029-5

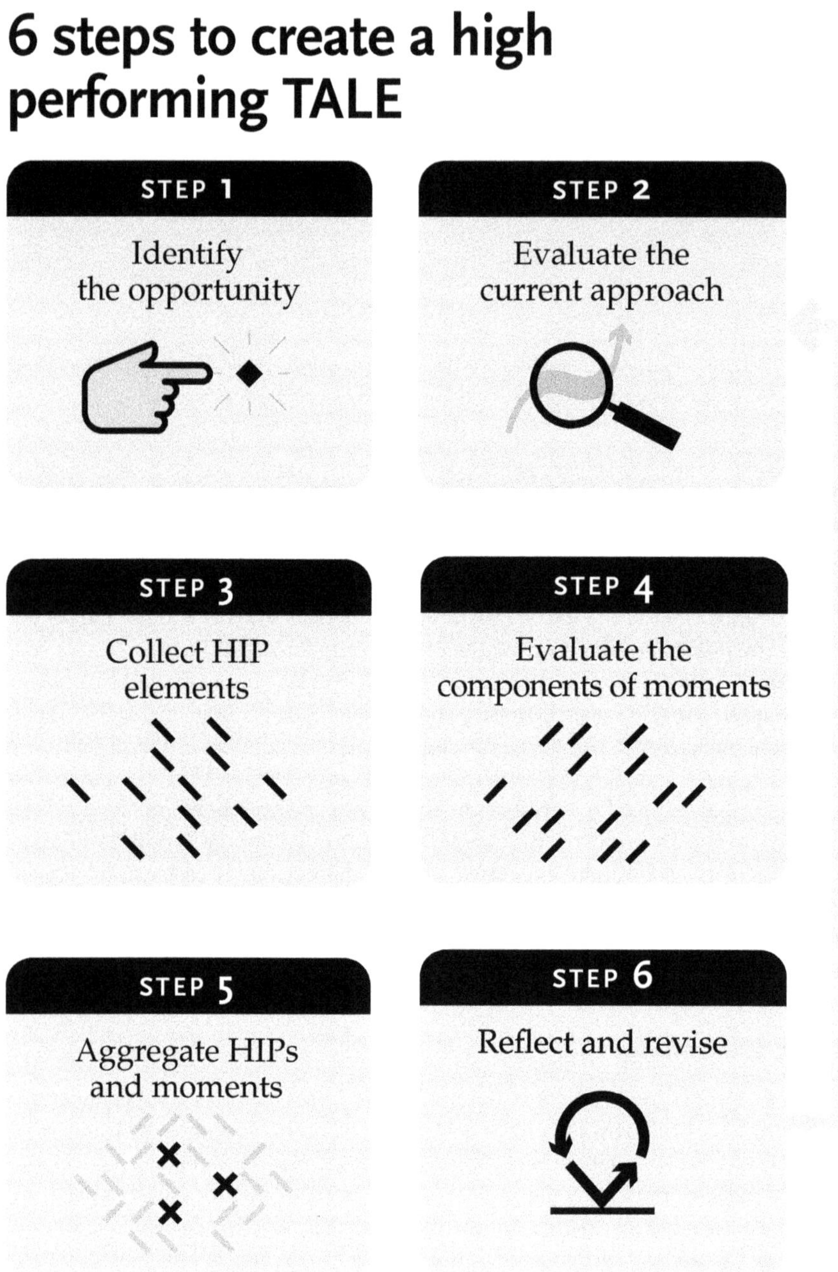

Figure 5.1 Graphical Process for Creating a TALE.

activity over multiple implementations. If a TALE doesn't go as planned, and you have added dozens of new parts to the activity, it may be difficult to know where to make amendments. Think of TALE creation like an experiment – isolate the variable you want to study.

STEP 1

Identify the opportunity

For many, identifying the opportunity for a TALE may be the easiest step because key educational aspects of classes often stand out – especially if you've taught the class before. Consider assessments, projects, or experiences that you currently have within the class. Remember, the best opportunities for creating a TALE are beginnings, times of transition, and endings.

If you're having trouble recognizing a TALE opportunity, consider the questions below to help identify points in the class that may be an opportunity:

- Are there opportunities for active, engaged learning?
- Where could students take responsibility for their own learning?
- How could the students connect with external stakeholders?
- What might create a memorable experience for students?

STEP 2

Evaluate the current approach

This step serves as a mechanism for you (the instructor) to evaluate how the identified opportunity relates to the overall context of the class and how it is currently delivered. This should be more of a big picture look rather than an exhaustive review (Figure 5.2). Greater detail will be highlighted as you complete steps three and four. Remember, if you don't have an existing approach that you're planning to modify, use this step to outline your initial approach.

Factor	Current Approach
At what time within the class does the TALE opportunity occur?	
What are the desired outcomes?	
What is the assessment?	
What is the instructional activity/approach?	

Figure 5.2 Evaluating the Current Approach.

STEP 3

Collect HIP elements

With the activity and approach identified, it's now time to explore the factors of a TALE. This begins with evaluating how HIP elements are currently included or can be included in the activity. The goal of this step is to identify any HIP elements that are being addressed in the current approach, then examine if there are opportunities for growth in the activity (Figure 5.3). Opportunities for improvement could occur in two forms: (1) improving the current approach where HIP elements are already being addressed or (2) adding activities to include HIP elements that are not currently being addressed.

HIP Element	Current Inclusion	Opportunity
Appropriately high-performance expectations		
Significant investment of time & effort by student		
Interactions with faculty and peers		
Exposure/Interaction with unfamiliar people and situations		
Frequent, timely, constructive feedback		
Periodic opportunities to reflect and integrate learning		
Discovery of relevance through real-world application		
Public demonstration of competence		

Figure 5.3 Evaluation of HIP Elements.

Remember, the goal is to make incremental improvements. That may mean leaving the "opportunity" column blank in one or more rows – no big deal. Oftentimes, seeing the activity occur in class will reveal opportunities that weren't recognized during development.

STEP 4

Evaluate components of Moments

Like the approach used above for the HIP elements, you want to consider what is currently being done and what could be done to improve the inclusion of Moments. Remember, when evaluating for Moments, the approach is broken down into the four components, Elevation, Insight, Pride, and Connection.

Figures 5.4–5.7 address components of the Moment. First, consider how or if you currently include each of the approaches shown. Then, score your efforts as "weak," "average," or "excellent." "Weak" items would be those that are marginally addressed. "Average" might be those items included but with significant opportunity for improvement. "Excellent" would be the items where your current efforts are achieving the components at a high level (recognizing they could still be improved). For example, an activity focused on the means and methods of exterior building envelopes:

- A "weak" example might be just studying various building envelopes.
- An "average" example may include walking outside of the building in which students are located to observe and discuss its exterior envelope makeup.
- An "excellent" example might be visiting an actual project under construction and engaging with the contractor and architect on challenges of that particular enclosure.

Step 4, Component 1-Elevation: Experiences that rise above the ordinary. These elicit feelings of engagement, joy, surprise, and motivation.

	1) Identify how the area is currently included in your class.	*2) Rank your current inclusion as "weak"(W), "average"(A), or "excellent" (E) for each area shown.*			*3) Based on your current inclusion and rank, detail your opportunity for improvement. Prioritize weak and average inclusions.*
Elevation Areas	**Current Inclusion**	**W**	**A**	**E**	**Opportunity**
Boost sensory appeal (i.e. real world experiences)					
Raise the stakes (i.e. competition, presentation away from campus, formal deadline)					
Break the script (i.e. celebration, inviting outsiders for presentations, breaking the monotony of the usual class)					

Figure 5.4 Moments of Elevation.

Step 4, Component 2-Insight: Experiences that cause realizations and transformations that change our understanding of an issue. An epiphany.					
	1) Identify how the area is currently included in your class.	*2) Rank your current inclusion as "weak"(W), "average"(A), or "excellent" (E) for each area shown.*			*3) Based on your current inclusion and rank, detail your opportunity for improvement. Prioritize weak and average inclusions.*
Insight Areas	**Current Inclusion**	**W**	**A**	**E**	**Opportunity**
Trip over the truth (situations that cause students to realize things they did not understand)					
Stretch for insight (encouragement from others, feedback from instructor that promotes growth by student)					

Figure 5.5 Moments of Insight.

Step 4, Component 3-Pride: Experiences that commemorate achievement. Identify what is motivating and celebrate it.					
	1) Identify how the area is currently included in your class.	*2) Rank your current inclusion as "weak"(W), "average"(A), or "excellent" (E) for each area shown.*			*3) Based on your current inclusion and rank, detail your opportunity for improvement. Prioritize weak and average inclusions.*
Pride Areas	**Current Inclusion**	**W**	**A**	**E**	**Opportunity**
Recognize others (i.e. students appreciating work of others; faculty/ peers recognition of students)					
Multiply meaningful milestones (i.e. including outside entities multiple times as part of an assignment)					
Practice courage (i.e. provide multiple opportunities for exposure for student leading to full assignment)					

Figure 5.6 Moments of Pride.

Step 4, Component 4-Connection: Experiences that create shared meaning. People connect when they work together toward a meaningful end.

	1) Identify how the area is currently included in your class.	*2) Rank your current inclusion as "weak"(W), "average"(A), or "excellent" (E) for each area shown.*			*3) Based on your current inclusion and rank, detail your opportunity for improvement. Prioritize weak and average inclusions.*
Connection Areas	**Current Inclusion**	**W**	**A**	**E**	**Opportunity**
Share in accomplishments (i.e. students share one success and one challenge as they move toward a goal or deadline)					
Share in purposeful struggle (ie. students work together toward a common deadline)					
Connect work to meaning (i.e. solve problems relevant to the real world)					

Figure 5.7 Moments of Connection.

We suggest that you consider the grading of each element and attempt to have at least one area (two suggested) in the "Excellent" category to realize that component of a Moment. The idea of evaluating each element individually provides an opportunity to prioritize where focus should be placed for improvement. Once your current situation is understood and your focus areas are identified, note specific opportunities for improvement that you may consider. You may be amazed to find that a single activity may address multiple components.

STEP 5 Aggregate HIPs and Moments

Once you get to this point in the framework, you've likely spent a good deal of time thinking deeply and critically to develop your strategies for HIP elements and Moments. Those steps help you think about each part of the framework separately, but they can also cause you to lose sight of the bigger picture. How complete is the experience? We want to make sure the approach addresses HIP elements and components of Moments in an overall context. Charting the intersection of the two in this step helps you see the big picture of your approach. Be forewarned, when you complete this step, you are likely to find some holes in your approach, i.e., there are some intersections that are empty. If you do have holes, that is absolutely okay. This step will allow you to see them, and if you feel they are severe enough, then you can make some adjustments to fill those holes as necessary. However, it is likely that they are not severe enough to warrant worrying about in this initial attempt. Once you have implemented the approach and have an opportunity to reflect, you can make revisions. Figure 5.8 readdresses the HIP elements and challenges you to find intersections with components of Moments.

Now, take the results from Figure 5.8 and consider where components of Moments intersect HIP elements. An example is shown in Figure 5.9. This is visual representation of the overall TALE potential. The more shaded (and marked) boxes that occur, the higher the potential for a TALE. The greatest TALE potential would have at least one box shaded in all Moment components but may not have all HIP elements addressed. Your resulting matrix should be the baseline tool used to compare your current efforts against future iterations of the activity.

Reconsidering HIPs

(Copied from Step 3)			*(Post Step 4)*		
HIP Element	Current Inclusion	Opportunity	Overlap of Moments YES	NO	Specific Moments being Overlapped
Appropriately high performance expectations					
Significant investment of time & effort					
Interactions with faculty and peers					
Exposure/ interaction with unfamiliar people & situations					
Frequent, timely, constructive feedback					
Periodic opportunities to reflect and integrate learning					
Discovery of relevance through real-world application					
Public demonstration of competence					

Figure 5.8 Reconsidering HIP Elements.

TALE Strength Summary

Where do the HIPs and Moments intersect in your proposed plan?

Elements of HIPs	Components of MOMENTS: ELEVATION	INSIGHT	PRIDE	CONNECTION
What will you do to **set the expectations** so they are challenging for the students but not discouraging to the point they give up?	X	X	X	X
How will the experience be set up to require a **significant investment of time and effort** by the students?	X	X	X	X
How can the experience involve the students having **interaction with faculty and peers**?	X	X		X
What can be done to have students **interact with others** that they wouldn't normally interact with in the traditional class experience?	X		X	X
What aspects of the experience will **provide** students with **frequent, timely, and constructive feedback**?		X	X	
How can you incorporate **periodic opportunities within the experience for students to reflect** and integrate what they've learned?		X	X	
How can you make the experience feel more **real-world** for the students instead of a ficticious learning exercise?	X	X		
What can you do to have the **students demonstrate their knowledge and skills** to others outside the class (public display)?	X		X	

X *Indicates an area where HIP element and Moment align*

Figure 5.9 Matrix Used to Consider Intersection of HIP Elements and Components of Moments.

STEP 6

Reflect and revise

As with any continuous improvement process, it should always include a mechanism for reflection. This includes considering what went well, areas requiring improvement, and what should be removed altogether. There are numerous mechanisms for executing this part of the process, from surveying the participants to evaluating student performance metrics, or simply taking stock of the experience from your own perspective as the teacher. The more you can incorporate feedback from the different perspectives involved in the experience, the better your ability to make revisions for the future. This is also a good time to consider implementing some of those ideas you identified when completing the previous steps but decided to hold for the future. While reflecting, keep a few things in mind:

- You do not have to fix all identified weak points at once.
- You do not have to add all your "ideas for the future" in the next iteration.
- If there are no major weaknesses identified, you might want to try running the approach again with another class and then consider potential changes.

Figure 5.10 provides a format for reflection and revision.

Good Points

____

____

____

Bad/Weak Points

____

____

____

Proposed Improvements

____

____

____

Figure 5.10 Reflect on the TALE and Consider Revisions.

TALEs to Be Told

Up to this point we have walked through the framework to develop and improve activities to set the stage for a TALE. The next chapter provides an example of how to maximize the potential for a TALE in a class of less than 50 students detailing out each of the six steps. The remaining chapters address other specific types of classroom situations. Remember, the goal of the book is to create TALEs in any environment for any class type. We will do this by providing examples, breaking down the TALEs, identifying and recommending active learning strategies, and providing helpful tips along the way. Since education is a "practice," the chapters also explore some "crash and burn" examples where attempts at creating a student experience fall short of what was intended.

In a perfect world, all the subsequent chapters would be structured the same and the content would follow a nice sequence, but that isn't how teaching works. Different class types have different needs, structure, and approaches. The pertinent information for one type of class may be completely irrelevant in another. Because of this, the following chapters are loosely based on an outline, but are largely individual cases. If you want to read all the chapters in a row, that works. After reading the next chapter (Chapter 6), if you want to jump to the classes that are applicable to you, that works too. Regardless of how you choose to consume the remainder of this book, it is our hope that these strategies and examples encourage and strengthen your resolve to seek TALEs in your own classes.

Appendix on Educational Foundations

The fundamentals we discussed in this chapter use the backwards design approach to developing curriculum, which was popularized by the work of Grant, Wiggins, & Jay McTighe in their approach "Understanding by Design," and L. Dee Fink's "Self-Directed Guide to Designing Courses for Significant Learning." This design approach has become a staple in effective course structure. While each approach has unique characteristics, the overarching goal of both is to design learning experiences with specific intention to achieve identified learning objectives.

Determine the destination you are trying to reach

You've likely heard the expression "begin with the end in mind" made popular by Steven Covey's book, *The Seven Habits of Highly Effective People*. Determining your end goals for your class is the starting point of the backwards design process. The fundamental idea here is that you need to first identify what your students need to know and be able to achieve those goals by the time they complete the course. These are the objectives/goals/outcomes (for our purpose, these words mean the same) included in your syllabus. These are more than just a piece of the syllabus though; they should be considered strategically crafted statements that help clarify to all course participants the targets the class is trying to hit. Thinking ahead to building a TALE, creating your course objectives is a great first opportunity to establish appropriately high expectations for the learners.

Identify the evidence of reaching the destination

If the word "destination" makes you think of a travel experience, then "reaching your destination" is an obvious endpoint. Unfortunately, in the learning process, reaching our learning goal destinations is much more obscure. That is why we need to think critically about the types of evidence that determine when we have reached our goals. In most cases, assessments serve as this metric. When developing assessments for a class, it helps to think macro (major summative assessments) down to micro (minor low-stakes formative assessments). Refer to Figure 5.11 to help put this into context.

Let's summarize some key points about assessments:

- Summative assessments are used to evaluate achievement at a specific milestone point, such as at the end of a learning unit, midterm, or end of semester. These are usually higher-stakes assessments where the students have minimal opportunity to correct gaps in their knowledge or skill.
- Formative assessments are opportunities for the instructor to evaluate the student's performance and give them feedback so they can adjust their performance before the summative assessment. Given their purpose, the more you

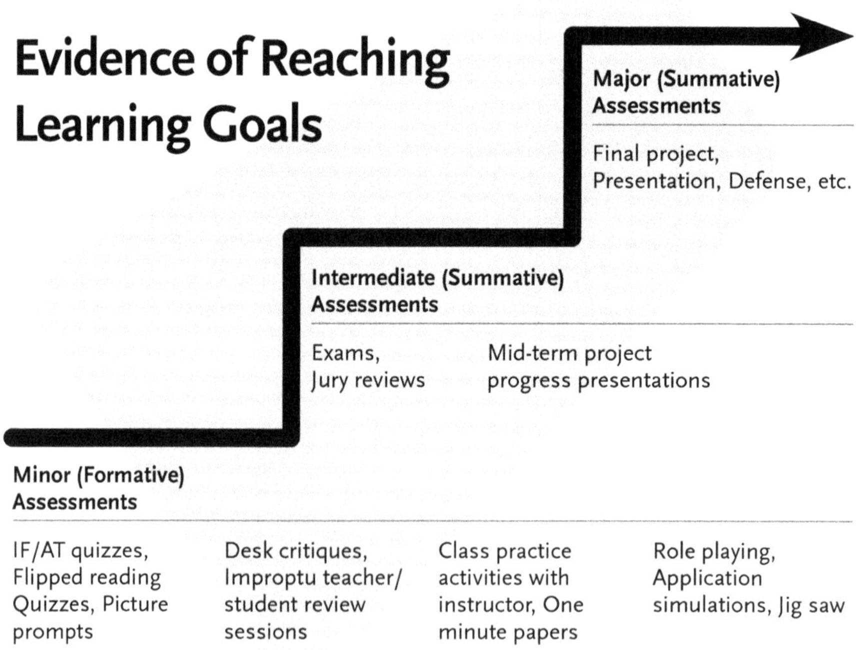

Figure 5.11 Formative Assessments Build to Summative Assessments.

can include formative assessments into the learning experience, the better opportunity your students have to improve.

- Do not assume that a lot of assessments mean a lot of grading. Assessments do not always have to count for a grade – especially formative assessments. Yes, grades are a motivating tool for students to do work, but you can encourage your students to do nongraded exercises by explaining their purpose and intended benefits.
- Craft "authentic" measurements (i.e., activities, performances, tasks, etc. that are reflective of what the students would be expected to do in their career). For example, if you are teaching students about the submittal process for a construction project, you might ask them to complete a review of a submittal. The student could even utilize a submittal log complete with transmittal form, then forwarding it to the next person for specific action. Not only is this far more engaging than a multiple-choice quiz, but it is also a more valid type of assessment that demonstrates the type of knowledge and skills the students would be expected to use in their careers.

Establish the route to the destination

Once the evidence of reaching the destination has been defined, you can now begin to plan the route for getting to the identified destination. This is where you develop the learning experiences to help the students build the knowledge and skills they will need to successfully reach the destination. The assessments will be critical points along the journey. Thought must be given to identify roadblocks students might encounter along the journey that prevents them from getting to those destinations. These identified roadblocks are where you need to provide instruction and consider adding pieces of HIP elements and components of Moments to elevate the learning experience or attempting a TALE. The submittal process assessment described above provides a good example. Considering roadblocks the students might hit when trying to execute that task will help identify instructional strategies to eliminate those issues. For example, the students may not know the exact order of operations nor specific individuals involved in the submittal process. One option to clarify this could involve an exercise where the steps in the process are printed on a set of index cards. Students may work in groups during class to outline the correct steps. Based on their input, a class discussion could provide feedback, clarifying and resolving misunderstandings. This gives the class an opportunity for immediate feedback and makes the learning more memorable and easily retrievable when they execute the submittal assessment.

6 Putting TALEs into Practice

What has been your experience with class size and your most effective classes? Can you connect the two? What opportunities afforded in small and mid-sized classes lend themselves to creating a TALE?

The previous chapter walked through the framework for creating a TALE. In this chapter, we will implement the six steps of the TALE framework by applying it to a mid-sized class scenario. In addition to the application example, this chapter will also serve to introduce the remaining chapters of the book that focus on TALE examples within specific class and content types.

The Case Considered

Our first example looks at creating a TALE in what we're classifying as a "mid-sized" class of 30–50 students. The class in this case is a construction estimating class taught by one of the authors.

Clarification moment: if estimating is not your area of expertise, don't worry. You should still be able to follow the logic presented here and apply it to the classes you teach. You will likely find areas more relevant to your field of expertise in the subsequent chapters.

The example estimating class is part of a construction management curriculum. It is the first and only major estimating class students take in the curriculum, and it follows a course on reading construction documents. The course covers aspects of material classification and quantification, pricing strategies, organizing and assembling estimates, and reviewing subcontractor pricing quotes (bids). Over the semester, students complete assignments and tests that assess their development of these competencies. The class concludes with a major activity as a summative assessment designed from the established course learning objectives. This assessment – called "bid day" – simulates a real-world estimating scenario

DOI: 10.1201/9781003106029-6

that many contractors execute when trying to win a project. Note: If you are not familiar with summative assessments or course learning objectives, refer to the appendix on educational foundations in Chapter 5.

Students spend the semester creating an estimate from a given set of project plans, and the semester culminates with the "bid day" simulation that occurs on the last class day. In the simulation, students receive about 80 different bid proposals over a three-day period from various fictitious subcontractors. They must review all the bids, compare the subcontractor numbers against their estimated cost, determine the most appropriate bidder for each work area, and adjust their overall project estimate accordingly. Once they have compiled all the bid information and updated their project estimate, the students submit their final bid for the project. Just as in a real-world scenario, the students must submit their bids by a specific day and time in sealed envelopes, with the prescribed bid form. If they submit anything incorrectly, their bid is considered nonresponsive (consequences for being nonresponsive are point deductions on the assignment score). The simulation ends with the instructor conducting a public bid opening with all the students and identifying the "winning bidder." The bid opening does not determine final scores on the bid day assessment; evaluation of the submissions is done by the instructor days after the class is over using a rubric. While the existing assessment has some active learning, HIP elements, and components of Moments, examining it through the TALE framework may look quite different.

STEP 1 Identify the opportunity

The end-of-semester "bid day" assessment was seen as an opportunity for a TALE:

- Scheduled at the end of the semester
- Already has active learning elements

STEP 2 Evaluate the current approach

- Natural summative assessment for class
- Recognized room for improvement

In Figure 6.1, the current approach to "bid day" has been evaluated through a variety of factors.

Factor	Current Approach
At what time within the class does the TALE opportunity occur?	The activity occurs during the last two weeks of the semester as a summative assessment.
What are the desired outcomes?	Students demonstrate the ability to organize a project estimate into appropriate component parts; analyze sub-contractor estimates; formulate a complete project estimate; expose students to the "hard bid" approach to selecting a contractor for a project.
What is the assessment?	Students complete an estimate that has been organized into work packages (component parts), has subcontractor bids added and evaluated, and a completed project bid form submitted per the owner's identified requirements.
What is the instructional activity/approach?	The teacher spends two class periods explaining the concept of the "hard bid" contractor selection approach, organizing a project estimate in preparation for receiving subcontractor bids, and presenting an overview on how to evaluate subcontractor bids.

Figure 6.1 Evaluation of the Opportunity for a TALE.

STEP 3

Collect HIP elements

HIP Element	Current Inclusion	Opportunity
Appropriately high-performance expectations	Students process 77 subcontractor bids over a two-day period to develop an estimate that is closest to the teachers estimate.	
Significant investment of time & Effort by Student	This happens over 1.5 weeks, so there really is no significant investment of time currently.	
Interactions with faculty and peers	No. The only interaction currently is with the teacher.	
Exposure/interaction with unfamiliar people and situations	Definitely not because the only current exposure/interaction is between the teacher and classmates.	
Frequent, timely, constructive feedback	The first real feedback they get is at the end when the entire activity is finished when the instructor presents the solution. But they don't get any specific feedback about their performance until after the semester is over when the instructor reports the grades for the activity. At that point they have no opportunity to talk with the instructor about their performance.	
Periodic opportunities to reflect and integrate learning	No. The timeline of activity currently is too compressed, and it's at the end of the semester, so there is no opportunity for the students to reflect and integrate learning.	
Discovery of relevance through real-world application	The teacher explains that the activity mimics a real "bid day" scenario that a contractor would go through, but it doesn't mean that the students actually discover the relevance.	
Public demonstration of competence	Student "bid" submissions are posted for the entire class to see, but they are anonymous so it's not much of a public demonstration.	

Figure 6.2 Evaluation of Current Inclusion of HIP Elements.

As we saw in the previous chapter, the chart for completing this step includes a column to identify current HIP element inclusions and a column to identify an opportunity for improvement. To help with clarity in illustrating the thought process for completing step three, we have broken it into two parts. The first part shows Figure 6.2 completed to identify how HIP elements are currently being included without addressing opportunities for improvement (our next part). In the estimating class, three of the eight HIP elements have been identified as currently included.

The second part of step three elaborates on identifying opportunities for improvement. As you go through the exercise of identifying current areas of inclusion, it is likely you will identify weaknesses in the approach and have ideas for improvement. Add these ideas in the opportunity for improvement

HIP Element	Current Inclusion	Opportunity
Appropriately high-performance expectations	Students process 77 subcontractor bids over a two-day period to develop an estimate that is closest to the teachers estimate.	Keep current approach—seems to be an appropriate level of challenge for most students.
Significant investment of time & effort	This happens over 1.5 weeks, so there really is no significant investment of time currently.	Break the experience up over the course of the semester so it is less compressed.
Interactions with faculty and peers	No. The only interaction currently is with the teacher.	Nothing identified
Exposure/interaction with unfamiliar people and situations	Definitely not because the only current exposure/interaction is between the teacher and classmates.	Include a general contractor to explain how they approach this type of activity.
Frequent, timely, constructive feedback	The first real feedback they get is at the end when the entire activity is finished, and the instructor presents the solution. But they don't get any specific feedback about their performance until after the semester is over when the instructor reports the grades for the activity. At that point they have no opportunity to talk with the instructor about their performance.	Add feedback points on organizing estimate into work packages, analyzing subcontractor bids, and inputting bids into the overall estimate.
Periodic opportunities to reflect and integrate learning	No,the timeline of activity currently is too compressed, and it's at the end of the semester, so there is no opportunity for the students to reflect and integrate learning.	Nothing identified
Discovery of relevance through real-world application	The teacher explains that the activity mimics a real "bid day" scenario that a contractor would go through, but it doesn't mean that the students actually discover the relevance.	Keep current simulation, but add a presentation from a general contractor about how they approach the process.
Public demonstration of competence	Student "bid" submissions are posted for the entire class to see, but they are anonymous so it's not much of a public demonstration.	Keep current approach.

Figure 6.3 Opportunities of Enhancement for HIP Elements.

column. Although we have broken them up into separate figures for clarity in this example, filling in both columns simultaneously as you go is fine. Not all opportunities for inclusion are going to come easy. There may be some HIP elements that require a bit more thought than others, so you will need to revisit them. As you can see from Figure 6.3, we were not able to identify an opportunity to include "Interactions with Faculty and Peers." If you get stuck with one, don't spend a lot of time pondering it. Move on through the rest of the steps and consider coming back to it later. It is not required that all HIP elements be addressed; in fact, we recommend making incremental changes over multiple semesters.

STEP 4

Evaluate components of Moments

Component 1 – *Elevation*

The current approach incorporates a real set of project plans, and the bid day setup adds elements of realism to the student experience that help make it somewhat memorable. But this is still a fictional scenario for the students. There is little clarification by the instructor that the project is part of a much bigger experience. For the students, it is just a set of drawings of some project that they are going to estimate over the course of the semester. The instructor also serves as the owner, interpreter of the project documents, general contractor, and subcontractor, which really discounts the inherent diversity of players present in the process. In addition, all the subcontractor bids being reviewed are fabricated by the instructor using fictitious contractors. Figure 6.4 provides a summary of how we might elevate the existing experience.

To improve the experience, we are adding a real owner element by working with the facilities management department at our institution to select a building project they have already completed, serving as the project for the semester. During week one, the key individuals that worked on the project will come to class and present the project to the students. This presentation will be framed as the "bid announcement" for the project, and the students will be working as "employees" of the general contractor bidding on the project. By making these changes, we are adding experience with diverse groups and creating relevance to real-world applications. Items like the "bid announcement" introduce the process as part of a larger picture and set high expectations for the class. The delivery approach elevates the traditional approach in a way that will make it more memorable for the students. It also sets a foundation for heightened performance expectations, deeper connections with others in the experience, and pride in what they will be doing.

Step 4, Component 1-Elevation: Experiences that rise above the ordinary. These elicit feelings of engagement, joy, surprise, and motivation.					
	1) Identify how the area is currently included in your class.	*2) Rank your current inclusion as "weak"(W), "average"(A), or "excellent" (E) for each area shown.*			*3) Based on your current inclusion and rank, detail your opportunity for improvement. Prioritize weak and average inclusions.*
Elevation Areas	**Current Inclusion**	**W**	**A**	**E**	**Opportunity**
Boost sensory appeal (i.e. real world experiences)	Simulation mimics a real "bid day", students use real drawings			X	Maintain current approach
Raise the stakes (i.e. competition, presentation away from campus, formal deadline)	Hard submission time with consequences for submitting late		X		Students work in teams submit their "bids" at a firm date & time at a real owner location instead of in a classroom. Teams will compete for submitting most accurate bid.
Break the script (i.e. celebration, inviting outsiders for presentations, breaking the monotony of the usual class)	Simulation conducted in a different classroom		X		A real owner will present the project to the students at the beginning of the semester, and the owner will conduct the bid opening at the end of the semester.

Figure 6.4 Current Approach and Opportunities for Elevation.

Component 2 – *Insight*

A key element of the "bid day" assignment is the actual bid day simulation. This is the culminating event of the experience that mimics a real bid day situation for a contractor. Out of the roughly 80 subcontractor bids that students receive over three days, about 50 of those bids are delivered to them over the last hours of the simulation. The students are expected to review and add them to the appropriate section of their estimate. As you might imagine, this is a stressful experience for most students.

When introducing the assignment, the instructor explains what the simulation is like, but few students really grasp the intensity of the scenario. The importance of being prepared and organized for the simulation is emphasized as well. It isn't until the students experience the simulation that most really grasp the full intensity.

> "That was the most stressful intense thing I've ever done in school."
> "That was horrible."
> "We definitely were not prepared for that."

At the point that most realize this, it's too late because this realization comes at the end of the simulation. While this is a profound moment for most of them, it is often perceived by the students in a negative sense rather than a positive experience. When looking at the idea of creating insightful Moments, the goal is to make this a more positive experience.

Our revised model will keep the original bid day simulation but also include a "practice" simulation a few days before the real simulation. This will allow students an earlier opportunity to "trip over the truth" of the bid day expectations. This early realization could suppress their motivation to continue preparing for the full simulation by thinking it is too difficult. To counteract this, a feedback loop will be added after the practice simulation to provide encouragement to the students that they can do it if they prepare appropriately. The feedback loop will incorporate the teacher and a general contractor facilitating a class session to help the students better understand how to prepare for the simulation and correct issues from the practice simulation. A presentation and practice activity will also be delivered by the general contractor on how to set up documents. The students then review their documents with the contractor and teacher for feedback. Execution of the full simulation will then provide an opportunity for them to discover that with preparation and understanding, they truly are capable. Figure 6.5 provides a summary of how we might enhance insight.

Step 4, Component 2-Insight: Experiences that cause realizations and transformations that change our understanding of an issue. An epiphany.

	1) Identify how the area is currently included in your class.	*2) Rank your current inclusion as "weak"(W), "average"(A), or "excellent" (E) for each area shown.*			*3) Based on your current inclusion and rank, detail your opportunity for improvement. Prioritize weak and average inclusions.*
Insight Areas	**Current Inclusion**	**W**	**A**	**E**	**Opportunity**
Trip over the truth (situations that cause students to realize things they did not understand)	Actual Bid Day experience—Students realize the importance of good organization in preparing for bid day in order to be successful, but it doesn't happen until end of exercise.	X			Add a "practice" a couple weeks before the simulation to give students exposure to the true intensity of the full-scale simulation before the full-scale bid day simulation.
Stretch for insight (encouragement from others, feedback from instructor that promotes growth by student)	Nothing in the current approach	X			Add a feedback session from the instructor after completing the "practice" simulation to clarify where students can improve and provide encouragement that they can succeed.

Figure 6.5 Current Approach and Opportunities for Insight.

Component 3 – *Pride*

When analyzing the current approach, pride was either going to happen for a student or it wasn't. The most likely opportunity for pride was if the students' submitted bid was close to the target bid, but there was no guarantee that would occur or even if that would yield pride. In many cases, the current approach created feelings of anger, frustration, and failure – but not pride. The revised approach will look to correct this by including a recognition ceremony and a general contractor. The ceremony will serve to recognize the hard work of all students. In addition to the general contractor, an owner will be included, and both parties will facilitate the recognition ceremony. Their acknowledgment of the student's accomplishments will help increase the opportunity to develop pride. Figure 6.6 provides a summary of the approach to creating pride for students.

Step 4, Component 3-Pride: Experiences that commemorate achievement. Identify what is motivating and celebrate it.					
	1) Identify how the area is currently included in your class.	*2) Rank your current inclusion as "weak"(W), "average"(A), or "excellent" (E) for each area shown.*			*3) Based on your current inclusion and rank, detail your opportunity for improvement. Prioritize weak and average inclusions.*
Pride Areas	Current Inclusion	W	A	E	Opportunity
Recognize others (i.e. students appreciating work of others; faculty/ peers recognition of students)	Nothing in current approach.	X			A dinner will be held for all students with the owner, general contractor, and instructor, and include a recognition ceremony acknowledging the winning team and the great work of all teams.
Multiply meaningful milestones (i.e. including outside entities multiple times as part of an assignment)	Nothing in the current approach.	X			Have a contractor do a presentation on how they approach these types of bids, have students do a practice setup of a bid tab, and have the contractor review it with them.
Practice courage (i.e. provide multiple opportunities for exposure for student leading to full assignment)	Nothing in the current approach.	X			The "practice" simulation will address this area; students will review their practice bid tab sheet with the contractor.

Figure 6.6 Current Approach and Opportunities for Pride.

Component 4 – *Connection*

Connections in this context include classmates as well as other individuals such as teachers, mentors, outsiders, etc. The whole idea is to provide an opportunity where the students unite with others to work toward a common goal. If executed correctly, students feel they are sharing the experience with someone and have something in common with those individuals. It is likely that some of the strategies we have included in the HIP elements and Moment components one through three will address this area. Some options to address connections include working in teams, interacting with outside professionals on specific tasks, and creating learning teams within the course (i.e., if you assign roles to different students in the class, create opportunities for students to break out into groups of their roles and talk with each other about their experiences).

In the current bid day setup, the students work individually, but there is opportunity for them to share in similar struggles and connect work to meaning. However, there is no formal mechanism included in the current setup that encourages students to interact with each other to find those connections. If it happens, then it is purely through student initiative to interact and discuss what they are experiencing. It is likely this will happen to some extent, but there are no guarantees – especially for the more introverted student.

The revised approach helps facilitate connections in two ways: (1) working in teams creates a natural opportunity for them to connect on the experience through their struggles and accomplishments, and (2) the inclusion of the owner and general contractor affirming to the students that what they're experiencing is realistic, adds opportunity for the students to connect their work to real-world meaning. This element is much stronger than the current approach where the simulation is isolated to the classroom with no outside involvement. Figure 6.7 provides a summary of how we might create a stronger connection.

STEP 5

Aggregate HIPs and Moments

After evaluating our HIP elements and components of Moments separately, we then went back and reviewed the approach and aggregated the results. Figure 6.8 shows the summary of the approach and where aspects of completing step four further included HIP elements in the approach. As you'll see, there are areas of significant intersection between HIP elements and components of Moments.

When completing this step for our estimating class, you can see from Figure 6.9 that we have developed some intersections between HIP elements and components of Moments. You will also notice that we have not filled all the boxes. Remember, the goal is not to try and fill all the boxes. In fact, it is unrealistic to think that every box is going to get shaded, and this is not the expectation. The empty boxes weren't anything of major concern; however, they did help us identify some areas of potential opportunity. Some approaches just are not going to be

Step 4, Component 4-Connection: Experiences that create shared meaning. People connect when they work together toward a meaningful end.					
	1) Identify how the area is currently included in your class.	*2) Rank your current inclusion as "weak"(W), "average"(A), or "excellent" (E) for each area shown.*			*3) Based on your current inclusion and rank, detail your opportunity for improvement. Prioritize weak and average inclusions.*
Connection Areas	Current Inclusion	W	A	E	Opportunity
Share in Accomplishments (i.e. students share one success and one challenge as they move toward a goal or deadline)	Completing the bid simulation	X			Include the recognition ceremony where the owner and general contractor acknowledge the students' achievements and what they accomplished
Share in Purposeful Struggle (ie. students work together toward a common deadline)	Working through the setup process for the bid day simulation—however no structured discussion/interaction among the students is provided.	X			Have students work in teams to complete the bid day simulation.
Connect Work to Meaning (i.e. solve problems relevant to the real world)	Nothing in the current approach	X			General contractor does a presentation explaining that what the students went through is what they actually go through in a real bid day situation.

Figure 6.7 Current Approach and Opportunities for Connection.

appropriate for some intersections, and that is okay. The goals of this step are to get a picture of what you have accomplished, identify noticeable areas of concern, and detail areas for future consideration. The intersection table is extremely helpful when it is time to reflect and revise.

STEP 6

Reflect and revise

When thinking about this step of the process for our estimating class, the first thing was to determine what information to use. The teacher's experience implementing the activity is an obvious data point. Performance metrics of the students is another pretty clear data point. But the perspectives of the others participating in the experience is also key. To capture these data points, we used a survey questionnaire to collect feedback from the students and held a debrief meeting with the general contractor and owner. This returned several great pieces of feedback for reflection and revision. Figure 6.10 highlights some of the key points.

Reconsidering HIPs

(Copied from Step 3)			*(Specific Moments being Identified from the Component Tables of Step 4)*				
HIP Element	**Current Inclusion**	**Opportunity**	Elevation	Insight	Pride	Connection	
Appropriately high performance expectations	Students process 77 subcontractor bids over a two-day period to develop an estimate that is closest to the teachers estimate.	Keep current approach—seems to be an appropriate level of challenge for most students.	X	X	X	X	Also add a real owner and general contractor
Significant investment of time & effort	This happens over 1.5 weeks, so there really is no significant investment of time currently.	Break the experience up over the course of the semester so it is less compressed.	X	X	X	X	Real owner presents activity at beginning of semester; add practice simulation a couple weeks before real bid day; general contractor does a presentation on how they setup and approach a bid day and work with students to develop preparation materials.
Interactions with faculty and peers	No. The only interaction currently is with the teacher.	Nothing identified	X	X		X	Students work in groups; instructor and general contractor give feedback.
Exposure/interaction with unfamiliar people and situations	Definitely not because the only current exposure/interaction is between the teacher & classmates.	Include a general contractor to explain how they approach this type of activity.	X		X	X	Include a real owner.
Frequent, timely, constructive feedback	The first real feedback they get is at the end when the entire activity is finished, and the instructor presents the solution. But they don't get any specific feedback about their performance until after the semester is over when the instructor reports the grades for the activity. At that point they have no opportunity to talk with the instructor about their performance.	Add feedback points on organizing estimate into work packages, analyzing subcontractor bids, and inputting bids into the overall estimate.		X	X		Practice simulation with feedback provided by instructor afterward; feedback from general contractor about how students are setting up their preparation materials
Periodic opportunities to reflect and integrate learning	No, the timeline of activity currently is too compressed, and it's at the end of the semester so there is no opportunity for the students to reflect and integrate learning.	Nothing identified		X	X		Practice simulation activity; general contractor feedback on setting up preparation materials
Discovery of relevance through real-world application	The teacher explains that the activity mimics a real "bid day" scenario that a contractor would go through, but it doesn't mean that the students actually discover the relevance.	Keep current simulation, but add a presentation from a general contractor about how they approach the process.	X	X			Inclusion of owner facilitating the simulation and explaining how their real process matches the simulation
Public demonstration of competence	Student "bid" submissions are posted for the entire class to see, but they are anonymous so it's not much of a public demonstration.	Keep current approach.	X		X		Bid submission and bid opening all done at owner's location instead of classroom

Figure 6.8 Reconsidering HIP Elements.

TALE Strength Summary

Where do the HIPs and Moments intersect in your proposed plan?

Elements of HIPs	ELEVATION	INSIGHT	PRIDE	CONNECTION
	Components of MOMENTS			
What will you do to **set the expectations** so they are challenging for the students but not discouraging to the point they give up?	X	X	X	X
How will the experience be set up to require a **significant investment of time and effort** by the students?	X	X	X	X
How can the experience involve the students having **interaction with faculty and peers**?	X	X		X
What can be done to have students **interact with others** that they wouldn't normally interact with in the traditional class experience?	X		X	X
What aspects of the experience will **provide** students with **frequent, timely, and constructive feedback**?		X	X	
How can you incorporate **periodic opportunities within the experience for students to reflect** and integrate what they've learned?		X	X	
How can you make the experience feel more **real-world** for the students instead of a ficticious learning exercise?	X	X		
What can you do to have the **students demonstrate their knowledge and skills** to others outside the class (public display)?	X		X	

X *Indicates an area where HIP element and Moment align*

Figure 6.9 Intersection of HIP Elements and Components of Moments for the TALE.

Good Points
Students loved inclusion of the real owner and general contractor.
Having the bid opening off-site at the owner location was a big hit with students that "added a lot of realism".
Spreading the experience out over a longer period of time allowed the students opportunity to reflect on what they were learning and be more prepared for the actual bid day simulation.
The general contractor really enjoyed the opportunity to connect with the students and "share their experiences with the students".

Bad/Weak Points
The practice simulation was good, but students said it wasn't extensive enough for them to get the full idea.
The general contractor really enjoyed being involved but would have preferred a project that they had worked on so they were more familiar with it and could offer other resources of information that would be more relatable.
Students like the inclusion of the real owner location for the submission of bids and bid opening, but they didn't like how many of them could not be part of the experience submitting the bid and bid opening because they were working on creating the final estimate in the classroom.
Inclusion of the real owner and GC was great, but it took a tremendous amount of time and coordination. The GC also mentioned this but didn't indicate that it would prevent them from doing it again.

Proposed Improvements
Instead of making the practice simulation 10 minutes with 3 bids, make it 30 minutes with 15 bids.
Work with the contractor and owner to find a project they worked on together and use that as the project for the class.
Incorporate a live stream camera in the bid room so students can watch the live bid room feed to see what's going on there.
Nothing was mentioned by anyone involved, but as the teacher, it seemed like there is opportunity to make the subcontractor bid component more realistic. Need to explore options for incorporating real sub bids in the future.

Figure 6.10 Reflecting on Implementation.

Conclusion

The example course started with a good base model built on sound educational theory using the backwards design process and active learning. However, there were opportunities for improvement, and the estimating class example illustrated an approach to building a TALE into a class. As is shown in Figure 6.10, we were able to incorporate several elements to create a TALE, but the reflection also indicates that our proposed modifications to the estimating class still have opportunities for potential improvement. Remember, this is perfectly acceptable. Your approach does not have to be perfect in the first attempt. Taking incremental steps to improve the experience provides tangible value to the students. And, who knows, perhaps after completing an evaluation of your potential TALE, you feel some pride and connection through the insight of elevating your students' learning experience.

7 Large Classes

> Engaging 100+ students in a lecture hall is complex but not impossible. Are there TALEs or pedagogical successes you had in a small or medium size class that could be extrapolated to a larger class?

In a perfect world, all our classes would be small, diverse, and have an even number of students. For most of us at medium to large universities, that is not how course sections play out, especially those teaching introductory classes. Many universities across the United States have classes where the number of students greatly exceeds the number of instructors. In some cases, this ratio could be 100+ students to 1 instructor. Based on what you learned earlier in the book, this may seem like a tough place for a Transformational Active Learning Experience (TALE). But with careful planning, a willingness to try (and possibly fail in front of hundreds of students), and an incremental approach to class improvement, large lecture classes do not have to be devoid of TALEs.

The Basics

Large classes are taught in, you guessed it, large lecture halls. Often these classes have rows of chairs in "theater style" seating with small or no desk space for laptops or note-taking. The chairs are facing giant screens the instructor uses to convey their message. Within these spaces, sound, lighting, student interaction, and the distraction potential for students are often problematic. At a certain point, the total number of students being taught within a space becomes irrelevant; whether you are teaching 80 students or 400, many of the same problems remain.

What Are the Benefits?

The obvious benefit to large lecture classrooms is the number of students who can take a single section of a class. This benefit certainly feels more like a perk to a department or the university, but a true benefit to the faculty does exist, and it lies within the large class population. A class with 200 students will have hundreds of diverse experiences, opinions, and approaches. Fostering a culture that can turn

DOI: 10.1201/9781003106029-7

that potential energy into kinetic energy is what this chapter is all about. Collaborative groups within these large classes can often present innovative ideas and viewpoints to other classmates. Bonds that can form within such groups foster new friendships and study partners that extend well beyond any in-class activity.

Are There Drawbacks?

Challenges exist when teaching large classes (Figure 7.1).

Challenges Faculty Face in Teaching Large Classes:

Adopted from Cuseo (2007) and Prosser & Trigwell (2014)

- Increases faculty reliance on the lecture method of instruction
- Reduces students' level of active involvement in the learning process
- Reduces the frequency and quality of instructor interaction and feedback to students
- Reduce students' depth of thinking inside the classroom
- Limits the breadth and depth of course objectives, course assignment, and course-related learning outside the classroom
- Lowers students' academic achievement (learning) and academic performance (grades)

Figure 7.1 Selected Challenges in Large Classes (Cuseo, 2007; Prosser & Trigwell, 2014).

The issues from papers by Cuseo and Prosser & Trigwell are certainly an obstacle but implementing strategies toward purposeful development of a TALE could mitigate or eliminate these challenges.

Representative Active Learning Strategies for Large Classes

Active learning strategies can be used as a tool to mitigate some of the drawbacks with large classes. As illustrated in Chapter 4, there are numerous strategies to consider. We have added this section to the remaining chapters to highlight specific strategies we feel are ideal for the given class type. For large classes, the active learning strategies in Figure 7.2 can help students engage at a deeper level.

Regardless of the active learning strategies deployed in large lecture classes, remember two things. First, many students will be uncomfortable voicing opinions to a large class of peers. When attempting discussions, consider using polling software for quantitative and qualitative responses anonymously. If you are

Representative Active Learning Strategies for Large Classes

Student self reflection/"One minute paper"

Students reflect on what they've learned by authoring an approximately 1-minute response to a question posed by the teacher.

EXAMPLE: You have identified in your Structures 1 class that students should be able to explain the difference between a moment and shear force. During your lecture, you embed the question, "Explain the difference between a moment and shear force" and give the students two minutes to write a response to the question. After two minutes, present the correct answer and then hold class discussion about any misunderstandings that may exist.

Think-pair-share

Teacher poses a question to the class; students formulate an individual response; and pair up with someone in their proximity to discuss their responses. Critical thinking is fostered. Often think-pair-share will lead to group discussion followed by whole class discussion.

EXAMPLE: While discussing material properties of different concrete assemblies in your materials and methods class, ask students to identify the different materials for an interior concrete slab-on-grade and explain the general purpose of each material in groups of two. In lieu of asking students to voice their opinions, you use a software like Google Docs, Box Notes, or Padlet for students to share their results with the entire class instantaneously.

Peer review/peer evaluation/group evaluation

Similar to think-pair-share, peer review allows students to participate in a low-stakes review of work while engaging in higher-level cognitive skills by analyzing an evaluating the work of others.

EXAMPLE: To help students review for an upcoming test, you develop a mini-practice test students complete on an individual basis. When done, students pair up and "grade" each other's practice test. The teacher then goes through the answers with the students to confirm correct answers. You may also choose to incorporate a software to manage the process and anonymize the students (if this is a concern).

Large group discussion

In groups of 10 or more, students discuss questions presented by the teacher in regards to a reading, video, or topic introduced in class.

EXAMPLE: As part of your class module on working in collaborative teams you post a video on the class website that presents a scenario on conflict resolution. You ask the students to watch it before the next class, and during the next class you put them in large groups to discuss a list of questions you've presented them about the video. The questions focus on the students evaluating conflicts and resolution strategies illustrated in the video.

Figure 7.2 Selected Active Learning Strategies That May Work Well in Large Classes.

worried about inappropriate questions or comments, the software allows you to review the content before it is posted. Second, student feedback in large classes is important. In small to medium-sized classes, faculty can often spot issues of cognition, retention, or concern from "reading the room"; in large classes, this is almost impossible. Get feedback on in-class activities and assessments from the students through an anonymous Q&A or comments session. This activity can be time-consuming, but the benefits to both the students and the faculty are immense. Student opinions are being heard, often improving the overall culture of the class, while faculty simultaneously benefit from the commentary.

Utilizing active learning strategies can be a great starting place for elevating your class. In the remaining portions of the chapter, we will present a TALE for large classes followed by an attempted TALE that did not go so well – a crash and burn.

The TALE

Professor Kidd teaches a first semester Environmental Controls class in the second year of the Architecture curriculum. It is the students' first experience in a class with all 80 sophomores studying Architecture, piled into a large theater-style lecture hall. Professor Kidd has built the class around a series of topics that impact the environmental conditions within their classroom building which include the following:

- Heat transfer/psychrometrics
- Thermal comfort
- Climate/microclimate
- Solar geometry and control
- The exterior envelope regarding heat and moisture transfer
- Passive heating and cooling
- Daylighting

Roughly two weeks are spent on each of the above topics. Through lectures, active learning exercises, small homework projects, and assigned work, students gain knowledge about each topic. After each two-week period, Professor Kidd assigns a culminating case study on each concept that connects the material learned about the topic to a real-world situation. Groups of three students are created at the beginning of the semester and remain the same for the entire semester. To develop these groups, Professor Kidd meets with the students' first-year studio professors, working with them to get the groups assigned by criteria. Professor Kidd's goal is to have groups with mixed academic performance, skill level, and time management. This system is not perfect, and can lead to problematic groups, but often the diverse skill sets within each group yield excellent results. For each case study, the groups craft a formal presentation that responds to the issues within the case study. Professor Kidd reviews each of the responses to questions posed and then selects specific groups to present each answer to the class. For example, group one may present the solution to question one while groups five and six present the solution to question two. During these presentations, Professor Kidd is mediating the discussion, and clarifying vague or incorrect responses.

In comparison with other classes that just dive right into group assignments, Professor Kidd uses the first lesson on heat transfer to really prepare the students for an excellent group experience. Before assigning the first case, a test is given on the material. The students take the test individually and submit it; then, Professor Kidd puts them in their groups, and they take the same test. This gives students an opportunity to begin working together, while simultaneously leading to rich discussion on heat transfer. Both test scores count toward the grade, giving accountability on both individual efforts and teamwork. The process yields strong preparation for the case study, and the group test provides an opportunity for the students to begin developing an approach to solving challenging problems together.

The following comments on instructor evaluations identify some items that helped build this TALE:

> "Our small group discussion on real-world issues really engaged me in the class. In other classes, I am not sure how to apply the concepts I'm learning. In this class, I know."
>
> "The group work was awesome. My group really gave me the confidence that I can solve complex, challenging problems."
>
> "I loved the class discussions on the case studies. It gave me immediate feedback on the work our group had done. And, when I did not understand a particular response, Professor Kidd was there and available to give an immediate answer."
>
> "The case studies were hard. At least I have a better idea when to speak up and when to be quiet when working in groups."
>
> "The group work was great. I learned how to listen more closely and how to craft a counterargument. I especially liked how our group had such diverse backgrounds which really provided new perspective to how I viewed some of the issues."

Connecting HIPs and Moments in the TALE

Using the TALE Strength Summary in Figure 7.3, we can evaluate the intersection of the elements of High-Impact Educational Practices (HIPs) and components of Moments. This can illuminate why Professor Kidd's approach was so impactful for the students.

Elevation

Professor Kidd "raises the stakes" by asking the students to demonstrate their knowledge to their colleagues through the group test and the formal presentation.

TALE Strength Summary

Where do the HIPs and Moments intersect in your proposed plan?

	Components of MOMENTS			
Elements of HIPs	ELEVATION	INSIGHT	PRIDE	CONNECTION
What will you do to **set the expectations** so they are challenging for the students but not discouraging to the point they give up?		X		
How will the experience be set up to require a **significant investment of time and effort** by the students?		X		
How can the experience involve the students having **interaction with faculty and peers**?				X
What can be done to have students **interact with others** that they wouldn't normally interact with in the traditional class experience?				
What aspects of the experience will **provide** students with **frequent, timely, and constructive feedback**?			X	
How can you incorporate **periodic opportunities within the experience for students to reflect** and integrate what they've learned?				
How can you make the experience feel more **real-world** for the students instead of a ficticious learning exercise?		X		
What can you do to have the **students demonstrate their knowledge and skills** to others outside the class (public display)?	X			

X *Indicates an area where HIP element and Moment align*

Figure 7.3 Intersection of HIP Elements and Components of Moments for the TALE.

Insight

Students "stretch for insight" by solving small problems, giving them confidence to address the bigger challenges. In addition, the high expectations, the significant investment of time to prepare the case responses, and the real-world examples used, coupled with consistent feedback and class discussions create an opportunity for students to "trip over the truth."

Pride

Using multiple case studies throughout the semester allows the students to celebrate and reflect on the completion of each finished case. Professor Kidd is "multiplying the milestones" while simultaneously having the students "practice courage" through the formal presentations.

Connection

Students "create a shared meaning" through group work with colleagues and the feedback from Professor Kidd.

Crash and Burn

You walk into the lecture hall, there are 150 students looking your way, and you are about to attempt a TALE that you have never done before… what could go wrong? The panic, fear, second-guessing your activity design – this is all very normal. The truth is, it could go fantastically, completely fall apart, or (most likely) will go over reasonably well but will need some tweaks. Professor Rushton's experience is one such example.

Professor Rushton's introductory engineering course, focused on statics, was taught in a lecture hall of 100 students. The class grade was based solely on three tests and a final. It was the first really challenging course in the curriculum, averaging around a 30% failure rate each semester. Professor Rushton was eager to improve the success rate of his class. He was willing to help the students in any way, but ultimately wanted each student to build study habits and become cognizant of their learning. Not only did he want them to be successful in his class, but he also wanted them to thrive in the curriculum. He wanted to develop student metacognition.

Professor Rushton worked to modify his class in a way to encourage students to take responsibility for their own learning – something he had identified when reviewing research on improving student performance. A key part of his efforts included offering an optional extra class session one night during the first week of the semester to teach students how to learn. During this session, he taught the traditional study method of "preview, attend, review, study, assess" – an approach he had read in a research publication by Dr. Saundra McGuire. When executed correctly, the method provides a platform for students to be strong lifelong learners and allows them to take responsibility for their learning. In addition, Professor Rushton focused on making sure students could parse out what instructors wanted them to learn in various classes, and how classes have structure and course objectives. The few students that attended the session departed appreciating some of the ideas but wondered what it had to do with the introductory engineering course and if they would ever have an opportunity to apply the ideas in a class.

During the class before the first test, Professor Rushton took the last few minutes to make sure all students had heard the "preview, attend, review, study, assess" mantra, not just the students that attended the extra class during the first week. Despite all his efforts to help the students, the failure rate did not change. Professor Rushton was frustrated and spent much of the semester wondering why students had not embraced responsibility for their own learning. Students continued to struggle on remaining tests, just like in previous semesters. Professor Rushton wondered if he would ever successfully engage today's generation with owning their own learning.

How Might This Be Changed to Provide a TALE?

Let's consider a few items Professor Rushton might have done differently that could change the experience for students and ultimately get them closer to an awareness of their learning.

- There was no attempt to identify "learning how to learn" (metacognition) as an outcome of the class. By setting expectations that focus on the student's future, a visible opportunity for a student-centered, forward-looking approach was missed.
- By offering the session on learning strategies only once during the first week of the course, the students couldn't understand why they needed the information. What if these sessions were offered multiple times – soon after major exams or assignments within the course??
- There was no assignment in the class that encouraged the students to practice and reflect on the experience – remember the only assessments were the three tests and the final. With no other graded assignments, the learning strategies taught in the extra class session were missing key elements of feedback prior to the summative assessments.
- Professor Rushton did not foster student discussion on the learning strategies. With limited opportunities for students to understand the value of the learning strategies from other students, the impact of the approach was limited to a few students.

Conclusion

The TALE provided by Professor Kidd is a great example of how some careful planning can dramatically change the experience for students. Although the illustrated example might be considered a light TALE due to the minimal intersections it contains, it demonstrates a transformational experience for the students in a large class that is historically difficult for faculty. The Crash and Burn illuminates that not all TALE attempts are successful, especially when done for the first time. With a little improvement and adjustment, Professor Rushton would not only have a positive experience but one that has the potential to shift how each student understands their learning. A skill that is invaluable for life.

References

Cuseo, J. (2007). The empirical case against large class size: Adverse effects on the teaching, learning, and retention of first-year students. *Journal of Faculty Development, 21*, 5–21.

Prosser, M., & Trigwell, K. (2014). Qualitative variation in approaches to university teaching and learning in large first-year classes. *Higher Education, 67*, 783–795. https://doi.org/10.1007/s10734-013-9690-0

8 Service-Learning

Many public institutions have a mission to meet the needs of the community. Imagine a time in your life where you were able to use your expertise to help someone. How did it make you feel? What intangible benefits did you get from the experience?

Students who participate in learning beyond the classroom have opportunities to contribute to community well-being and simultaneously enrich their educational experience. By emphasizing concepts like partnerships, shared creation of projects, and mutual respect among all participants, faculty can elevate student service-learning experiences.

The Basics

Service-learning combines community service for societal needs with a pragmatic, progressive educational experience. Students are provided the opportunity to directly apply classroom knowledge, bridging the divide between theory and practice. When done well, service-learning benefits the community and reinforces academic knowledge gained in the classroom.

Service-learning can have a variety of scopes. It could include anything from a weekend project to a semester-long, credit-bearing class. As applied disciplines, those in the Architecture, Engineering, and Construction (AEC) education spectrum are uniquely poised to engage in such practice.

What Are the Benefits?

In 1990, Ernest Boyer suggested the expansion of scholarly work within higher education to meet the needs of the community. "Application" was one of Boyer's four pillars of engagement.

As this "application" occurs, students benefit in a wide variety of ways:

- Hands-on use of skills and knowledge that builds upon relevant academics
- Interaction with people of diverse backgrounds

DOI: 10.1201/9781003106029-8

- Increased problem-solving skills and social development, often leading to heightened self-efficacy
- Increased civic responsibility and a commitment to service
- Enhanced leadership and communication skills

The built environment is as much about people as it is technical knowledge, and service-learning is a powerful tool to train empathetic and culturally sensitive professionals. In Figure 8.1, the benefits to students, faculty, and the community are illuminated.

Student Benefits from Service-Learning

- Incorporates hands-on use of skills and knowledge that builds upon relevant academics (Makes classroom knowledge real)
- Offers interaction with people of diverse backgrounds
- Increases problem-solving skills and social development (which often includes hightened self-efficacy)
- Increases civic responsibility and a commitment to service
- Enhances leadership and communication skills

Faculty Benefits from Service-Learning

- Generates new teaching ideas
- Matches many universities, especially land-grants, commitment to betterment of community.
- Application of research and teaching outside of the classroom
- Service and/or outreach required for many tenured and tenure-track professors

Community Benefits from Service-Learning

- Meets basic human needs
- Develops synergies between university, the community, and community partners
- Fosters opportunity for community members to engage with university students
- Supports caring for others

Figure 8.1 Benefits of Service-Learning.

Are There Drawbacks?

Whether you examine service-learning from a student, faculty, or community perspective, there are challenges. Finding good community partners and spending time to prepare the work is difficult, and projects that do not meet shared student, faculty, and community goals can lead to frustration. Logistically, taking a group of students outside of the university can be rife with red-tape and administrative requirements. Add in the need for materials, tools, safety training, personal protective equipment, undeveloped skill sets, and of course, money, and you have a challenge that sounds insurmountable for many faculty; especially when not supported fully by institutions and departments. Even simple projects done over a week's timeframe could cost $5,000 or more. Expenses can be offset by nonprofits, endowments, and grants, but the acquisition of these funds is time-consuming and is required each time a service-learning project begins.

From the student side, attitude and circumstances can be obstacles. While most students approach service with enthusiasm, there are students who see these projects as a chore. In cases like this, defining meaningful outcomes can sometimes help. Beyond motivation, there are almost always students with specific circumstances that make their participation challenging. These could be complicated work or school schedules, transportation issues, family challenges, or a variety of other issues.

Representative Active Learning Strategies for Service-Learning

Service-learning will inherently include some active learning. The goal here is to identify active learning strategies you might be able to incorporate that will elevate the experience for the students and eliminate some of the drawbacks mentioned above. The list in Figure 8.2 is not exhaustive, but we hope you can find a strategy that you can apply.

The TALE

Each year Professor Changa offers a one semester, disaster relief service-learning elective. The class identifies a place where a natural disaster has occurred; connects with on-site partners; plans the work to be conducted; visits to aid in the rebuilding of the community for a week; and reflects on the experience. For Professor Changa, the logistical complexities of taking a trip to a disaster area pale in comparison to the experience and emotions that are tied to helping people who are in desperate need of assistance. Students can apply skills learned in the classroom, engage deeply with peers and those in the local community, and experience the satisfaction of helping people in need.

Running a service-learning class is not for everyone, especially in a disaster area. As Professor Changa says, "You have to be willing to check your pride at the door." Prior to the application phase of the class, the students are made aware that many of the comforts in their lives may be out of service in the disaster area. Students are met "head on" with the stark challenge of the conditions while

Representative Active Learning Strategies for Service-Learning

Self-assessment

Students complete a quiz (typically ungraded) to determine their competency level of the subject.

EXAMPLE: After students complete a service-learning project, have them author a written self-assessment of the experience. Questions such as "What worked well? What contributed to your learning?"; and "What one thing could be changed to improve the learning experience?" are great starters.

Experiential learning

Activities that allow students to observe and apply concepts discussed in class to a real-world situation.

EXAMPLE: Have students present in groups their collective experience with service learning. Consider inviting those outside of class that were engaged with the project.

Inquiry learning

Students use an investigative process to make observations, pose hypotheses, and speculate on conclusions about questions posed by the instructor. Then students share their thoughts and tie the activity back to the main idea/concept.

EXAMPLE: Define the final product students are to produce through the service-learning project. Find a benefactor or partner that will serve as a client in the service-learning project, providing a way for students to work through the client (third party) while you, as a professor, guide and coach only when required.

Figure 8.2 Selected Active Learning Strategies That May Work Well in Service-Learning.

given a realistic task (usually basic construction) and time frame (a week or less). This action sets a foundation of high expectations and is further developed by providing examples of previous service projects that have impacted communities (including speakers from those trips).

Generally, the group (including Professor Changa) sleep in a communal space, on army cots, in a church or gymnasium. Lunches are brown-bagged, and normal tasks, like doing laundry, are unlikely. Before leaving, the students create budgets, develop schedules, assign roles, and evaluate resources. Many of these endeavors fill the learning outcomes of the class, which require the application of skills and knowledge taught in the classroom to a real-world setting. During this preparation time, Professor Changa works with her partners, generally local nonprofits or ministries, to ensure a safe and effective trip for her students. In addition, Professor Changa prepares the students for what to expect, including the proper use of social media in a place often still reeling from the storm.

During the week of service-learning, the hands-on nature of the construction work is difficult, and the hours are long, but the impact for the students is

TALE Strength Summary

Where do the HIPs and Moments intersect in your proposed plan?

Elements of HIPs	Components of MOMENTS: ELEVATION	INSIGHT	PRIDE	CONNECTION
What will you do to **set the expectations** so they are challenging for the students but not discouraging to the point they give up?	X			X
How will the experience be set up to require a **significant investment of time and effort** by the students?		X		X
How can the experience involve the students having **interaction with faculty and peers**?		X		
What can be done to have students **interact with others** that they wouldn't normally interact with in the traditional class experience?	X	X		X
What aspects of the experience will **provide** students with **frequent, timely, and constructive feedback**?		X		
How can you incorporate **periodic opportunities within the experience for students to reflect** and integrate what they've learned?		X	X	
How can you make the experience feel more **real-world** for the students instead of a ficticious learning exercise?	X			X
What can you do to have the **students demonstrate their knowledge and skills** to others outside the class (public display)?	X	X	X	

X *Indicates an area where HIP element and Moment align*

Figure 8.3 Intersection of HIP Elements and Components of Moments for the TALE.

significant. By working side by side with those impacted by the disaster, students experience a diversity not found in the typical classroom, all the while receiving real-time feedback on their efforts. And the intense, short-term nature of the work mandates that stronger peer relationships develop.

Professor Changa requires a post-trip reflection from the students. In these reflections, students acknowledge the importance of using their technical skills for the scope of work they executed within the week, but perhaps more impactful are the people they met. Professor Changa tells the story of a home they were helping to rebuild for a woman whose house was destroyed during Hurricane Michael in Jackson County, Florida. The scope for the week included installing the exterior vinyl siding, installing/taping/mudding the drywall, and building a back deck. As the week came to a close, the owner of the home, an elderly woman, wanted to give back to the students and offered to make them lunch. Professor Changa, knowing the time and expense the woman would exhaust, kindly declined, but the woman would not accept "no" as an answer. After hours of preparation and a cost that may have been a full week's wages for the host, the woman fed the entire class. This act of connection through giving back was not lost on the students. A woman who had lost almost everything to a hurricane was willing to spend her limited resources to thank them. Professor Changa's class later wrote in their reflections about this lunch, an unforgettable experience, the culminating event of an experience whose individual elements combined to form a TALE (Figure 8.3).

Connecting HIPs and Moments in the TALE

Service-learning is a perfect application to connect many of the HIP elements and components of Moments. Let's examine some of the detailed intersections from the TALE.

Elevation

Professor Changa's approach to focusing on the human-centric, real-world situation addresses both "high expectations" and "real-world application." Her explanation of the approach and the plan creates a team atmosphere immediately, providing a natural connecting point for students. The entire experience boosts the student's "sensory appeal."

Insight

Students "stretch for insight" through 6 of the 8 HIP elements. Especially key to insight are the interaction with people (peers, faculty, and those they serve) as well as the post-trip reflection.

Pride

As the students depart the project and recognize the value of a job done well. Students see their work (and the work of their counterparts) at the end of the week. The pride from this is magnified as students reflect on the experience and "recognize others" for their contribution to the project.

Connection

Connection is found through the "deepened ties" and "shared meaning" of the trip. Much of this happens over the extended trip but is amplified in the final days of the work as students connect the real-world application of their learning to the people they are helping.

Crash and Burn

Professor Myles had a reoccurring class during which he required service-learning. In one particular semester, he partnered with a church to have a group of students construct a small concrete patio. The patio was adjacent to the portico of the church and was currently occupied by a landscaped area. Professor Myles took care to assure the church he would cover the cost of the concrete and any formwork required. He made sure students were trained on safety related to handling and placing concrete as well as the tools required to complete the work. Then, he left them to complete the work.

Unfortunately, the students decided to complete the work on the last possible weekend of the project. They met on Friday, removed the landscaping, prepared the area, and erected the formwork. A couple of team members decided to travel out of town at the last minute for the weekend leaving only three of the five team members to complete the work on Saturday. On Saturday morning when they arrived at the church for the concrete pour, they learned that a wedding was planned. The church staff and the wedding party were less than thrilled that the work was ongoing. In one area, they had to bury a drainpipe below the slab before concrete arrived. As they trenched a small area for the pipe on Saturday morning, they hit a water main causing water to shoot up in the air almost 30 feet. Imagine the excitement of the wedding party with a new water feature at the church and no running water for the event. The team was able to shut off the water and complete the concrete placement, but, because of their rush, the slope of the patio was off, leaving the church with a product that would collect water instead of shedding it. The team members realized the challenges associated with volunteer work and were sorry the experience ended with an upset wedding party and a project that failed to best serve the needs of those they attempted to help.

How Might This Be Changed to Provide a TALE?

Professor Myles missed an opportunity for insight through strong faculty and student interactions. There was no proposal developed for the project, limiting time available for feedback (like schedule or buried utility concerns), and there was no plan for a prepour review where issues like slope may have been identified. Each of these opportunities at critical milestones of the project could have enhanced insight for the team. In addition, these interactions could have elevated the experience for the team creating a significant, but manageable hurdle in the educational exercise.

By allowing the work to occur just in time, few expectations for success were developed and there was no opportunity for the students to celebrate their work with the larger community. If a formal presentation of their work had been included, an opportunity for public display and development of deeper connection with the class and owner would have been created.

How Might I Start a Simple Service-Learning Project with Limited Resources?

As we previously discussed, there can be many challenges to developing a service-learning experience. The TALE example represents a robust and complex service-learning experience requiring significant faculty time, effort, and funding, with an enormous potential for failure. The Crash and Burn, a much cheaper and logistically simpler project, ended up a much worse experience for the students and the community. To address this, we have included "Do's" and "Don'ts" to help simplify the process of developing a service-learning experience (Figure 8.4).

Conclusion

Giving students the opportunity to tangibly apply classroom acquired skills is a fundamental part of AEC education. Service-learning can be challenging, but the return on investment is massive. These projects not only have a high potential to create a TALE for students, but also for the faculty managing the project. Consider starting small at first and work your way toward larger experiences like Professor Changa's disaster-relief class.

How might I Start a Simple Service-Learning Project with Limited Resources?

DO THESE THINGS!

- ✓ Define learning outcomes.
- ✓ Set expectations for learning goals and any assessment.
- ✓ Seek projects close to campus accessible by carpool/shared rides.
- ✓ Foster authentic partnerships between universities and community partners.
- ✓ Collaboratively define scope with project partner.
- ✓ Encourage partners to provide raw materials and equipment.
- ✓ Assure students have a reasonable schedule and budget within resource constraints.
- ✓ Form groups of 5-10 students and have them jointly develop a proposal for the work.
- ✓ Assure students are trained on relevant safety issues and have PPE.
- ✓ Have a pre-visit tutorial on hand tools and machinery that will be on site (and other hazards).
- ✓ Have a pre-visit discussion on the use of social media, especially in disaster or low-income areas.
- ✓ Allow the students to interact with the "client" and users of the project on which they are working.
- ✓ Have students present their proposal to you and the client where feedback can be provided in real time.
- ✓ Serve as coach providing timely, relevant, and meaningful feedback when needed.
- ✓ Build in reflection time.
- ✓ Have students document project and consider a final presentation where students present what they accomplished and what they learned.

DON'T DO THESE THINGS!

- ✗ Be rigid in requirements. Flexibility is required as we mix academia with real-world partners.
- ✗ Require scope completion for full credit.
- ✗ Assume all students have the skills to execute the desired scope.
- ✗ Assume the students have used electrical hand tools.
- ✗ Focus too much on the project without focusing on the people impacted by the project.

Figure 8.4 Do's and Don'ts for Simple Service-Learning Projects.

9 Internships and Co-Ops

What work experience early in your career facilitated your learning?

You may have heard students say things like "I learned more in my three-month summer internship than I did in all my courses." While there are many valid reasons to argue that the statement is false, it does encapsulate the general idea of the potential benefits in completing a degree-focused work experience in an Architecture, Engineering, and Construction (AEC) profession. Students and the industry have long seen work experience as transactional – a path to a job. We believe that work experience in the AEC industry can be transformational but achieving this often requires engagement of the educational institution.

The Basics

Internships are experiences where students work for a real company within the discipline in which they plan to obtain their degree and focus their career. These work experiences are usually for a set time period – most commonly over a summer break. A Co-Op, on the other hand, is a more formal program where students often alternate semesters between work and school.

These work experiences can come in many forms:

- Students pursue work experiences without a formal requirement from the university.
- Students receive course credit for completion of the experience.
- The curriculum requires an internship (not for credit) prior to graduation.

Higher education, including AEC disciplines, often see internships as something students do in addition to school. Internships that are not part of the curriculum are divorced from the obligation of the faculty. If an internship is credit-bearing or is part of a graduation requirement, then there are usually more structured requirements. Credit-bearing internships will have established course learning outcomes, student required preplanning, deliverables (such as weekly and final reports), and an instructor to whom the student will report.

DOI: 10.1201/9781003106029-9

For the employer, structuring the work experience for the student can vary. Some companies will have well-developed internship programs that outline what interns will do, who they will work with, and how they will transition through the work experience. Other companies may not have formal structures in place and will instead base the internship approach on their current needs.

What Are the Benefits?

Literature on the benefits of work experiences is vast, and the list below highlights some of the most common.

- Connect what is learned in school with how it is applied in the "real world"
- Develop soft skills like communication and time management
- Provide an opportunity to understand the roles and responsibilities of different career fields and/or industry positions
- Obtain full-time employment out of their internship/co-op pools
- Build one's professional network
- Determine if their identified career is a good "fit" for them
- Improve overall academic performance (Binder et al., 2015)

Are There Drawbacks?

Despite the many potential benefits of work experiences, there can be downsides.

- Some employers view interns as employees used for completing menial tasks. Students may not develop new skills or build upon academic knowledge gained. This can range from serving as a company courier to an unskilled laborer on a jobsite sweeping floors.
- The value of the work experience can be diminished if students don't have time to reflect on what they have learned, integrate with what they have done in school, and publicly convey what they have done.
- Students will go into a work experience with the perception that it will guarantee them a job after school. This is often not the case, and students need to be aware of this reality before entering into the work experience.
- Work can be challenging to find in a marginal economy, in rural areas, or at those universities that lack a strong professional network in the major. This is especially problematic when the work experience is part of a graduation requirement.

Representative Active Learning Strategies for Internships and Co-Op Classes

The very nature of internships and co-ops that involve the educational institution will involve active learning. The goal is to identify opportunities to deepen the experience for the students and eliminate some of the drawbacks mentioned above (Figure 9.1).

Representative Active Learning Strategies for Internships & Co-Ops

Case studies

Use detailed accounts of situations that occur to people, projects, families, events, etc., that prompt students to connect what they've learned with real-world situations, actions, and consequences.

EXAMPLE: You assign students to write a case on a subcontractor coordination issue that occurred on a project during the internship and document what the issue was, who was involved, and how it was resolved.

Self-assessment

Students complete a quiz (typically ungraded) to determine their competency level of the subject.

EXAMPLE: At the end of the internship, you have students write a reflection paper that asks them to critically analyze their work experience and what they learned, how it relates to the courses they've already taken, how it will help them in their future courses, and any new realizations about the future profession.

Role playing

Students assume the role of a person or a position as part of an "act" that accurately demonstrates a situation and/or people's roles in those situations. Preparing for the "act" helps students enhance their understanding of the concepts and theories being discussed. Executing the "act" demonstrates their level of understanding. Feedback from the audience after the act provides immediate feedback to the performers.

EXAMPLE: Student and classmates act out an experience that occurred during their internship. This could be something they participated in or just witnessed.

Figure 9.1 Selected Active Learning Strategies That May Work Well in Internships and Co-Ops.

The TALE

Katelyn is at the end of her internship course, and she's working on the final deliverable for the course – a case study developed from her experience. While trying to identify the focus topic for her case, Katelyn reflects on her experience of the past summer. She recalls the last day when coworkers continuously approached her, telling her how much they enjoyed having her there and how much they would miss her. She recalls the send-off lunch they threw for her where everyone from the company attended – even the field superintendents. During lunch, they shared stories with one another.

About midway through her internship, Katelyn played a part in aiding the project superintendent to solve a coordination issue between the plumber and mason. It was a profound event for her because of the intensity of the situation. About ten workers were waiting for them to find a solution so they could get back to work. They had discussed coordination issues in her classes before, but she never really understood the true severity of it until that experience.

One week she even had the opportunity to run the jobsite for a week under the supervision of the superintendent. Although nervous, she had fun with it. She embraced it well and even recalled the owner calling on her last day of work to wish her well and tell her how much he enjoyed having her working on his project. During the 10-minute call, the owner expressed his deep gratitude for her hard work. He closed by telling her to please reach out to him if there was ever anything he could do to help her, and that he knew she had a bright career ahead.

The exit meeting with her bosses was a profound memory. She knew when she started the internship that there would be meetings with supervisors but did not realize her final exit interview would involve all the supervisors she had worked with and the owners of the company. During the two-hour meeting, they each talked about her work and performance. The summary of her experience and praise were amazing for Katelyn. Among many things they discussed, an event during an owner meeting with the project manager really stuck out. They were trying to get a decision from the owner on color selection for the glass panels of the building's curtainwall system. The owner had been delaying the decision for weeks, and she remembered how the project manager was so graceful in communicating the importance for the owner to decide so it would not delay the project. She quickly realized the importance of communication strategies they had discussed in her classes. As she thought more about her summer, many more of these examples crept into her mind.

Katelyn knew she learned a lot during the internship. The weekly reports she had to submit to her teacher provided frequent opportunities to reflect on what she was learning. At the start of the summer, she never imagined that it would turn out to be such an impactful experience, but Professor Azham, the internship faculty, did.

Prior to the onset of the internship, he was strategically working with the employer to develop the internship experience. He made sure the employer embedded a diverse array of experiences that tied directly with the courses she had completed, and the ones she would be taking when returning to school. Some items included attending coordination meetings, facilitating jobsite management, and working in various areas of project management. Professor Azham knew there was opportunity to increase the impact for students by encouraging critical thought. Incorporating the case study assignment and weekly reports helped the students think deeper about their experience. These purposeful efforts by Professor Azham set the foundation of opportunity for Katelyn's realized TALE (Figure 9.2).

TALE Strength Summary

Where do the HIPs and Moments intersect in your proposed plan?

Elements of HIPs	Components of MOMENTS: ELEVATION	INSIGHT	PRIDE	CONNECTION
What will you do to **set the expectations** so they are challenging for the students but not discouraging to the point they give up?			X	
How will the experience be set up to require a **significant investment of time and effort** by the students?		X	X	X
How can the experience involve the students having **interaction with faculty and peers**?			X	
What can be done to have students **interact with others** that they wouldn't normally interact with in the traditional class experience?	X	X	X	X
What aspects of the experience will **provide** students with **frequent, timely, and constructive feedback**?		X	X	
How can you incorporate **periodic opportunities within the experience for students to reflect** and integrate what they've learned?		X	X	
How can you make the experience feel more **real-world** for the students instead of a ficticious learning exercise?	X	X		X
What can you do to have the **students demonstrate their knowledge and skills** to others outside the class (public display)?	X		X	

X *Indicates an area where HIP element and Moment align*

Figure 9.2 Intersections of HIP Elements and Components of Moments for the TALE.

Connecting HIPs and Moments in the TALE

Elevation

Professor Azham's work prior to the internship elevated the experience especially through the opportunity for Katelyn to interact with people different from her. Whether tradespeople on the job, senior leadership, or the owner, that interaction on real-world projects, coupled with the opportunity for her to tangibly demonstrate what she had learned through the case study, elevated the experience.

Insight

Key to Katelyn's insight was the significant amount of time she invested with the firm over the summer. In addition, Dr. Azham's timely feedback on the weekly reflection papers provided opportunity for Katelyn to further consider how her experiences with a variety of people in a real-world setting have expanded her learning.

Pride

Most of the High-Impact Educational Practice (HIP) elements provided pride for Katelyn. Especially evident were the effort and time committed by her sponsoring company to give feedback, and the celebration event at the end of the summer.

Connection

Katelyn's work with people on the job site solving real problems provided key connection points. This experience galvanized her with the company and made her long for the next problem to solve.

Crash and Burn

Jordan couldn't wait to start his summer internship. He was part of the first internship course in his program, and it would be his first experience working for a construction firm focused on management rather than skilled trades. On his first day, he met the office staff and his supervisor, Ernie. Ernie informed Jordan that he would be spending most of the summer on a jobsite working with their superintendent, Roger, on a new elementary school project. They were currently installing foundations, site utilities, and getting ready to start structural steel. He was thrilled because he had never seen this type of work before and knew it was a great opportunity.

The next morning Jordan reported to the jobsite promptly at 7 a.m. Upon entering the office trailer, Jordan was confronted by Roger who promptly asked in his gruff voice, "Who are you?" Jordan introduced himself, confirmed the man was in fact Roger, and then explained how excited he was for this opportunity. To Jordan's surprise, Roger's only concern was that Jordan stay out of his way. Roger didn't want an intern on his job and didn't have time to teach some young college kid how to manage a project. Jordan wasn't happy with the response but took the comments lightly, thinking Roger might have been having a bad morning. The remainder of that week Jordan continued to arrive promptly at 7 a.m., and each day Roger would do his best to ignore Jordan, using comments like, "Don't bother me, I'm busy." or "Go find somewhere to hide so you're not in my way." Concerned that Roger's attitude wasn't going to change, Jordan contacted Ernie to express his concern. Ernie assured Jordan that he would talk to Roger.

The following Monday when Jordan arrived, Roger was waiting for him and said he had a job for Jordan. Unfortunately, cleaning mud off the building foundations so that the waterproofing could be installed, was not what Jordan was expecting, and certainly not what he was supposed to be doing. Although it was miserable work, and he couldn't wait to be done, Jordan never complained. He hoped that his hard work would gain Roger's respect and garner other management-related job responsibilities. Unfortunately, as each week went on, Roger had more menial tasks for Jordan to complete. Some even put Jordan in questionable safety situations.

By midway through the summer with nothing changing after multiple inquiries to Ernie and the professor for his internship course, with no foreseeable end in sight Jordan decided to leave the internship. He had considered just not showing up for work on Monday without notice but knew that was not honorable. Jordan felt a great sense of relief that he didn't have to come back, but he was frustrated that he had let Roger get the best of him. Worst of all, he felt like he had wasted one of his limited opportunities to learn from a practical, applied internship, and now must retake the internship course.

How Might This Be Changed to Provide a TALE?

Reviewing the issue that Jordan experienced, let's consider what could have been done differently.

- Involvement of a faculty member as a liaison with the construction company may have provided intervention to correct the issues Jordan was experiencing. Ongoing reflection by Jordan and feedback by the faculty during the internship would also have helped identify the issue early and would have provided a platform through which the discussion with the company could have occurred.
- Faculty involvement prior to the internship may have established better shared expectations of the internship and intended outcomes.
- The lack of opportunity to engage with the diversity inherent on a construction project was missed and could have been different with faculty involvement.

Conclusion

These work experiences can be immensely powerful shaping moments for students. Because of our connection between the major and profession, schools of higher education within the AEC industry are uniquely poised to employ the high-impact practice of work experiences. With a little forethought and planning, and a purposeful connection to the student's experience, internships or Co-Ops can be an impactful academic endeavor.

Reference

Binder, J. F., Baguley, T., Crook, C., & Miller, F. (2015). The academic value of internships: Benefits across disciplines and student backgrounds. *Contemporary Educational Psychology*, *41*, 71–82. https://doi.org/10.1016/j.cedpsych.2014.12.001

10 Capstone Experiences

> Did you have an educational experience where you were required to assemble many of the things you had learned into a single project? How did it impact you? If not, what would such an experience have looked like for your undergraduate education?

We, the authors, teach in a program where a capstone experience (called "thesis") has been required for every undergraduate student for over 70 years. Students view the thesis as a "rite of passage." And if you talk to almost any graduate, the first story they share from their undergraduate experience is related to their thesis. Students often see the thesis as the last requirement to be met, and faculty approach the thesis as the last opportunity to share knowledge and skills needed for professional practice.

Designed initially to help transition students to the "real world," today's thesis experience emphasizes student reflection, synthesis of learning across academic and professional experiences, and the quest for graduates to be self-starters in learning and work.

The Basics

Capstone experiences require the graduating student, either alone or in small groups, to complete an academic project that incorporates the discipline studied in the last semester/quarter of their undergraduate experience. The capstone is defined by the AAC&U as "culminating experiences" that "require students nearing the end of their college years to create a project of some sort that integrates and applies what they've learned" (*Capstones and Integrated Learning*, 2013). Often, faculty design this experience to be a summative assessment of the student's mastery of material in the discipline, or, said differently, a measure of how well the student meets the learning outcomes of the major. There is no one-size-fits-all design to a capstone experience. Four concepts are identified below, but we encourage you to think broadly about what a capstone experience may look like in your program.

DOI: 10.1201/9781003106029-10

Independent Study

Within some overall framework, students select their own project. An undergraduate research project is a typical example. Students meet with a sponsoring faculty member who serves as a mentor, coach, and ultimately, the person who assesses the work. Often, this type of project includes some type of culminating public display of competence, such as a presentation.

Course-Based Study

Another common design of capstone experience is built on a single academic course. The course is often built around the learning outcomes for the major, and students are asked to apply many of the things they have learned over their undergraduate experience. This often involves a common project or type of project to which the student responds. In the Architecture, Engineering, and Construction (AEC) industry, this can often be tied to specific practice in industry providing unique opportunities for a Transformational Active Learning Experience (TALE).

Another effective way to operate a course-based capstone experience is to select a focused topic of interest to the faculty member leading the course. The students work to develop specific elements around the broader topic with each student developing a written narrative on their specific element. Students work together to craft an introduction and a conclusion allowing assembly of the full document into a pseudo-book the class publishes at the end of the semester.

Experience-Based Study

A third approach is to tie the capstone to an experience, such as a service-learning or internship experience.

Portfolio-Based Study

The final approach involves student portfolios. E-portfolios, identified as a specific High-Impact Educational Practice (HIP), are often the delivery mechanism. Typically, a capstone experience built around a portfolio has students collect artifacts from their work and reflect on each. In some cases, this could be geared toward the learning outcomes of the major so that their work illustrates the student has met the outcomes. In some cases, the student may have built the portfolio over the course of their time as an undergraduate student, and the capstone course focuses on specific reflections or even a deeper research project built around some element of their work. In a project like this, the faculty member works with the student to identify artifacts needing revision and encourages self-reflection on the part of the student.

What Are the Benefits?

Capstones offer multiple benefits for students:

- Provides opportunity for students to demonstrate (and realize) accomplishing learning outcomes of the major
- Provides opportunity for independent or collaborative work, both of which prepare students for the AEC profession
- Fosters soft skills of communication, responsibility of being a self-learner, confidence, and time management

There are also some benefits to faculty and the departments that operate capstones. First, the capstone provides a common rallying point for faculty regarding program assessment. As capstones are completed and evaluated by a broad spectrum of faculty, strengths and weaknesses within the overall degree program are identified (measures program learning outcomes); discussion is fostered; and shifts can be made to improve the academic outcomes. In much the same way, other lower-level classes tend to evolve as faculty discuss "what will the student need to know and be able to do" when graduating from this program. Such programmatic reflection can be powerful within an academic unit.

Are There Drawbacks?

Kinzie (2013) noted concerns regarding capstone experiences in the areas of access, diversity, inclusion, and quality. Equity concerns tend to focus on design and implementation as well as availability to all students. At times, capstone experiences are not appropriately planned, causing the quality of the academic experience to be poor. Infrequent or unstructured formative feedback during the capstone is a common student complaint. This can be magnified when the capstone is not aligned with the overall program learning outcomes. Faculty also report challenges on mentoring and assessing capstone experiences. Many of these challenges revolve around the lack of support faculty have from the institution (financial, time, and other resources) in capstone implementations.

Representative Active Learning Strategies for Capstone Classes

While specific approaches to address drawbacks are largely a function of the type of capstone, active learning strategies can assist faculty in having students reflect, consider, and revise their work at critical stages in the process (Figure 10.1).

The TALE

Professor Shreve loved the architecture capstone class she taught. It was an odd mix of celebrating all the learning students had experienced in the program

Representative Active Learning Strategies for Capstone Experiences

Strip sequence

Students are provided steps of an ordered process on sheets of paper in random order, then asked to construct the appropriate sequence.

EXAMPLE: Consider an undergraduate research project on the failure of a specific structure. The faculty may outline various pieces of the research project including materials and methods, existing literature on the failure, interviews with professionals, etc... By ordering these events that led to the failure, the student gains insight into the logical order of the work, and an opportunity to discuss the process at a deep level is fostered.

One-minute paper

Students reflect on what they've learned by authoring an approximately 1-minute response to a question posed by the teacher.

EXAMPLE: In an e-portfolio reflection, a student became focused on a singular view of an issue. In short, the student was focused on the internal function of a space without concerns for the impact on the exterior of the building. By asking the student to list all other possible viewpoints/perspectives over a one-minute exercise, the student was encouraged to revise their reflection to include multiple perspectives on the issue.

Inquiry learning

Students use an investigative process to make observations, pose hypotheses, and speculate on conclusions about questions posed by the instructor. Then students share their thoughts and tie the activity back to the main idea/concept.

EXAMPLE: In a project-based capstone, students are asked to identify a specific element of their project and develop the workforce approach for completing that project element. Specifically, they are asked to consider from the subcontractor perspective what the specific steps are for completing the work, industry standards for installation, the types and number of workers needed, how long it will take, and how much it will cost to complete the piece of work.

Figure 10.1 Selected Active Learning Strategies That May Work Well in Capstones.

coupled with reflecting on and revising their portfolios amassed during their architectural studies. She began the class by outlining the outcomes students were required to demonstrate prior to graduation. Focusing on four realms of critical thinking; professional practice; integrating architectural solutions; and building practices, skills, and knowledge; there were 26 specific outcomes that students were asked to demonstrate in their portfolios. After reviewing each of them, she provided a rubric for students to study and consider as they began the class. Students left the class realizing it was up to them to maintain the high-performance levels set by previous graduates.

The assignment was relatively direct. Students were to assemble artifacts from all their classes and internships (at least one required) into a single e-portfolio.

Then, they were to provide a written reflection on each of the elements demonstrating all 26 learning outcomes for the major. In cases where the work students had produced did not fully demonstrate the outcome or where competing objectives were identified, students were asked to revise their original work product from the previous classes and convey their thought process as to how the exercise added to their understanding of the required learning outcomes.

Students worked side by side in the studio environment on the capstone. Professor Shreve's presence and daily engagement with each student set clear expectations that significant time was required by the students to reach expected performance levels. Each day, she had specific goals for her interactions with students, and she found a variety of unique ways to ask different questions and encourage students. Students recognized her commitment and spent long hours in the studio working on the capstone project.

Over the 15-week semester, students were required to make bi-weekly, informal presentations. Each described a reflection on which they were working, including revisions or adjustments they made because of their current knowledge or experiences. This provided Professor Shreve an opportunity to gauge student progress and provide timely feedback to enhance further reflection. Each presentation ended with a discussion on the skills demonstrated and opportunities for future development.

Beyond the regular student presentations and the desk critiques, Professor Shreve found other ways to create opportunities for critical reflection. In one case, she designed a "so what" exercise. Students were to describe an incident that added to their architectural education. This included the who, what, when, and where. Students were then required to make sense of the significance of the action. Finally, students were required to indicate how that experience shaped them, including key lessons gained. This reflection then became part of the overall e-portfolio.

The culminating event of the portfolio capstone was a presentation to an industry panel that included a minimum of five of the elements and reflections created. Students almost unanimously agreed that although the class was difficult, it was one of the most affirming and clarifying experiences of their undergraduate experience.

Connecting HIPs and Moments in the TALE

Let's see if we can pinpoint some of the factors that created such strong educational experiences (Figure 10.2).

Elevation

Professor Shreve leveraged the opportunity for moments of elevation by setting high expectations on day one, reinforcing those, and crafting unique interactions between students and faculty throughout the semester (with feedback).

TALE Strength Summary

Where do the HIPs and Moments intersect in your proposed plan?

Elements of HIPs	Components of MOMENTS: ELEVATION	INSIGHT	PRIDE	CONNECTION
What will you do to **set the expectations** so they are challenging for the students but not discouraging to the point they give up?	X			
How will the experience be set up to require a **significant investment of time and effort** by the students?	X			
How can the experience involve the students having **interaction with faculty and peers**?	X	X	X	X
What can be done to have students **interact with others** that they wouldn't normally interact with in the traditional class experience?		X	X	
What aspects of the experience will **provide** students with **frequent, timely, and constructive feedback**?	X	X		
How can you incorporate **periodic opportunities within the experience for students to reflect** and integrate what they've learned?	X	X	X	X
How can you make the experience feel more **real-world** for the students instead of a ficticious learning exercise?		X		
What can you do to have the **students demonstrate their knowledge and skills** to others outside the class (public display)?	X	X	X	X

X *Indicates an area where HIP element and Moment align*

Figure 10.2 Intersection of HIP Elements and Components of Moments for the TALE.

Insight

The purposeful engagement throughout the semester encouraged students to think deeply about what they had learned and move beyond just a superficial response. Key to this were the presentations, critiques, and "so what" exercises explored weekly.

Pride

Pride was realized by the students as they connected the learning outcomes required to be successful in their fields of study with what they had learned throughout school. This was fully realized through their final presentations to industry.

Connection

Professor Shreve's emphasis for students to include all aspects of their coursework and internships encouraged them to realize relationships that otherwise may have been missed. Consistent and regular feedback helped students connect what they had learned with real-world practice and how their individual experiences would impact them moving forward. The final presentations also provided opportunity for the students to share in struggles while preparing and presenting, and the accomplishment of completing.

Crash and Burn

Instructor Downing was approached by one of her favorite students, Jesse, about sponsoring a capstone project for him. Instructor Downing was extremely busy with her research and had a four-class load that semester, but she almost felt like she could not say no. Jesse had been an excellent student, and she was sure he would do a great job with minimal oversite. As a civil engineering student with a focus on-site development, Jesse proposed a service-learning experiential capstone where he would work with the local farm museum to imagine and design an RV park that might become a reality for the community. Instructor Downing agreed, and Jesse was excited about the opportunity.

As the semester began, Instructor Downing and Jesse met to review the project. They identified a preliminary list of deliverables, and Jesse was challenged to reflect on this, discuss with the community partner, and resubmit as a formal record of the work ahead. Two weeks transpired before the form was returned. Instructor Downing reviewed the list and was a little concerned about how generic some of the deliverables were. But she knew Jesse needed to move ahead and assumed he would "figure it out" along the way.

Both Instructor Downing and Jesse had a lot ongoing. When they did meet, only generic feedback was provided, and it did not significantly impact how Jesse approached the issue. Jesse was unsure how to address some items, so he did what he thought made sense without really investing significant time to investigate.

At the end of the semester, Jesse and Instructor Downing set a date and time for a final presentation of the work. They invited the community partner, and Jesse presented the work accomplished. Instructor Downing realized quickly that Jesse had not solved the key questions identified by the community partner, nor had he achieved some of the most significant learning outcomes of the course. The community partner was frustrated that many of their questions remained unanswered. Jesse realized he had disappointed both his faculty champion and his community partner. His capstone was much less than what he had originally hoped it would be, and wondered, as he was about to graduate, if he was adequately prepared. For Instructor Downing, the assessment of Jesse's submission was problematic. With vague objectives and a lack of metrics, how would she evaluate Jesse's work?

How Might This Be Changed to Provide a TALE?

Instructor Downing needed more focus on expectations, deliverables, and a commitment to progress markers and feedback. Key to these were the following:

- A clear list of deliverables with appropriate expectations of all parties should have been developed before the semester.
- Regular connections with Instructor Downing, students, faculty, and industry providing relevant and timely feedback were needed.
- Engagement of the community partner as a participant in the process would have provided a more diverse experience with the opportunity to develop pride and insight.

Conclusion

Capstones offer opportunity for graduating students to demonstrate achievement of program outcomes through the application of skills and knowledge gained throughout their undergraduate experiences. While they often include elements of active-learning, Capstones offer great opportunities to incorporate a TALE.

References

Capstones and Integrated Learning. (2013). [Peer Review Journal provides a quarterly briefing on emerging trends and key debates in undergraduate education.]. Association of American Colleges & Universities. https://www.aacu.org/peerreview/2013/fall

Kinzie, J. (2013). Taking stock of capstones and integrated learning. *Peer Review, 15*(4). https://www.aacu.org/publications-research/periodicals/taking-stock-capstones-and-integrative-learning

11 Team-Based Student Competitions

Have you ever been a member of an educational team? What was your experience, and how did it impact you as a student or teacher?

Competition plays a significant role in the built environment. Think for a moment about the competition that occurs daily in industry:

- Competing to win a project
- Creating a purposeful ever-lasting design
- Producing a superior project on time and on budget with high quality
- Striving for continuous improvement (next project better than the last)

In academia, and especially within the Architecture, Engineering, and Construction (AEC) sector, students find value through scenarios and problems that mimic real-world applications. Because significant problems in practice are addressed by groups of individuals, our chapter on competitions focuses on team-based student competitions. By combining the competition inherent within the industry and the real-world applications we strive to imitate, a team-based contest can provide an excellent opportunity for students.

The Basics

Competitions as a teaching tool can be traced as far back as the 1st century BC where Marcus Flaccus, a Roman scholar, introduced the principle of competition with his students by rewarding books to students with the best essay on different themes (Baden, 2006). Award-based prizes like Flaccus' approach have shown to be an effective tool in motivating some students to "up their game."

What Are the Benefits?

Beyond increasing student motivation, the opportunity for a significant real-world learning experience is enhanced. Students are provided an opportunity to explore important topics to which they feel a personal connection. Opportunities

DOI: 10.1201/9781003106029-11

increase for students to apply problem-solving skills in a context that is relevant to them and their profession. The inclusion of diverse teams and an emphasis on collaboration among team members further strengthen learning.

Are There Drawbacks?

Some suggest that competitions only work for high-achieving students. The argument is that only these types of students are drawn to the "competitive" aspect of the experience as a motivator to "up their game." Similarly, competitions have been shown to demotivate lower performing students because they see it as a place where their inadequacies will be put on display. Competition can also be negatively perceived by different cultures. In cases like these, students will tend to shy away from the opportunity rather than embrace it.

Representative Active Learning Strategies for Competitions

A variety of other active learning tools can also be used with competition teams (Figure 11.1).

The TALE

It's the final week of the school year, and Professor Sey's commercial construction team is getting together to celebrate their time together and look toward their upcoming graduation. While reflecting on their experiences together over the past three years, Annabeth comments that her experience as part of Professor Sey's team is one that really sticks out as a great moment during her time at school. Other team members agree and share stories about nervousness making formal presentations to industry professionals, long nights practicing and evaluating project scenarios, the intensity of the 12-hour pressure-cooker at the competition putting their proposal together, and the excitement of the closing awards banquet when they placed second in their category. While some remember different resounding moments of the experience, they all agree that it was one of their best school experiences because it was "so real to what they would be doing in their careers." Blake recalls how the competition experience "totally changed my perspective on my classes…I had a new appreciation for the things we were learning and how it was relevant to what I would be doing."

Professor Sey coaches this team every year in a regularly scheduled competition organized by a professional organization. He selects eight students for the team each year from a combination of students that self-identify as interested in competitions and those that he recognizes would benefit but not necessarily self-select. Students work together on-campus to prepare for the off-campus competition event for about three months.

The actual competition occurs over a two-day period with one day devoted to the team solving a real-world problem posed by industry and one day devoted to team presentations, feedback from sponsoring companies, and team awards.

Representative Active Learning Strategies for Student Competitions

Role playing

Students assume the role of a person or a position as part of an "act" that accurately demonstrates a situation and/or people's roles in those situations. Preparing for the "act" helps students enhance their understanding of the concepts and theories being discussed. Executing the "act" demonstrates their level of understanding. Feedback from the audience after the act provides immediate feedback to the performers.

EXAMPLE: Place students in specific roles that would be typical of those associated with the competition event. One person may be the lead designer where another may be the constructor. Use the opportunity to have students perform different roles than what they may know best or even expect.

Mini competition (games)

The instructor poses questions or deliverables, and students work on them in groups or individually. Responses are presented to the class for evaluation and discussion. Often, this can be competition based.

EXAMPLE: Imagine a problem that could be solved in a single night that may have multiple good solutions but only a few great ones. The mini competition could focus on a key element associated with the larger competition and help prepare for "game day".

Jigsaw discussions

A general topic is divided into smaller, interrelated pieces like a puzzle. Each member of a team is assigned to become an expert on a different piece of the topic. After each person has become an expert on their piece of the puzzle, they teach the other team members about that piece.

EXAMPLE: In a competition to build a concrete canoe, one member may be the materials expert, while another the construction expert, while another the boat expert. Make the team members jobs to teach the others in the group what they learned about the role they are playing and how it contributes to the team's overall competition goal.

Forum theater

Students are the audience and actors as in a theater. The situation can be played out multiple times with different students taking different roles and perspectives.

EXAMPLE: As part of studying the roles of team members in the design process you show students a video of examples of this, some of which are inaccurate. Then, you have groups of students brainstorm on what the appropriate roles and processes should be and act out possible solutions with the best team recognized.

Figure 11.1 Selected Active Learning Strategies That May Work Well in Competition Teams.

Teams compete for their school wearing university colors and logos while competing against traditional sports' rivals. The students take great pride in the experience and are passionate about providing the best practical solution to the problem using their classroom knowledge.

Over time, Professor Sey has identified two key outcomes for students on his team:

- Apply construction management knowledge and skills to construct a solution to the competition problem as a collaborative team.
- Create a compelling presentation to summarize the solution to the problem posed.

Professor Sey uses that competition as the assessment. He has essentially created the outcomes through the lens of the competition. Along the way, Professor Sey creates a series of opportunities for students to advance toward the ultimate outcomes:

- Students are asked to make formal individual presentations on themselves to the team.
- Films of previous teams are viewed with meaningful reflection opportunities for students on what worked well and what could be improved.
- The team works together in practice to solve mini problems that mimic the competition.
- Students present their solutions of mini problems to an audience of professionals assembled by Professor Sey, to receive constructive feedback along the way.

After the competition, Professor Sey coordinates a debrief including the team and the judges. He notes that this time is where some of the "highest learning" occurs. Finally, Professor Sey organizes a dinner for the student participants to further celebrate the experience and the work accomplished.

Connecting HIPs and Moments in the TALE

Let's explore the experience of Professor Sey's students a little more and see if we can pinpoint some of the factors that created such strong educational experiences (Figure 11.2).

Elevation

The competition, by definition, mimicked gameday. Competing against other universities away from campus, on a real-world project, was a natural elevator. Professor Sey further elevated by setting expectations of both learning and winning (past performance and value experienced by previous teams), and his engagement

TALE Strength Summary

Where do the HIPs and Moments intersect in your proposed plan?

Elements of HIPs	Components of MOMENTS: ELEVATION	INSIGHT	PRIDE	CONNECTION
What will you do to **set the expectations** so they are challenging for the students but not discouraging to the point they give up?	X		X	
How will the experience be set up to require a **significant investment of time and effort** by the students?		X	X	X
How can the experience involve the students having **interaction with faculty and peers**?	X	X	X	
What can be done to have students **interact with others** that they wouldn't normally interact with in the traditional class experience?	X	X	X	X
What aspects of the experience will **provide** students with **frequent, timely, and constructive feedback**?		X	X	
How can you incorporate **periodic opportunities within the experience for students to reflect** and integrate what they've learned?		X		X
How can you make the experience feel more **real-world** for the students instead of a ficticious learning exercise?	X	X		X
What can you do to have the **students demonstrate their knowledge and skills** to others outside the class (public display)?	X	X		X

X *Indicates an area where HIP element and Moment align*

Figure 11.2 Intersection of HIP Elements and Components of Moments for the TALE.

of industry professionals and actual projects along the journey created milestones of elevation that prepared the group for the ultimate competition.

Insight

The competition practice was set up to allow students to solve small problems (both individually and as a team) that gave them confidence to address bigger challenges. And Professor Sey provided consistent feedback in the process. This fostered deep interaction for the team with peers, faculty, and industry.

Pride

Again, the set-up of the competition with students competing for their university naturally created an environment where pride could be fostered. Professor Sey complemented this by consistently recognizing team members; building courage through conversations with industry; and creating milestones of success along the way. And he wrapped up the event with a celebration where all team members were recognized.

Connection

The team environment with strong faculty and industry support provided strong connection points for the students. These connections were realized through collaborative problem solving, trips to projects and company sites, the formal presentation, and the team dinner.

Crash and Burn

Timothy's experience on a student competition team illuminates how the experience can be much different and emphasizes drawbacks as opposed to benefits. He recalls:

It was about 7:50 p.m. We had 10 minutes to get our proposal submitted to the judges. We still had numerous pages left to print, and then get the proposal five floors down to the judge's location. We had been working on our proposal since 7 a.m. and nearly everyone on our team was at each other's throats. As the hours of the day ticked away, disagreements between our team members escalated. Within the last couple of hours, two of our teammates from our partner university, left and never came back. We didn't know what work they had left to complete, and they didn't give us any of the stuff they had been working on. The three of us that were left started blaming each other for why we weren't going to make the 8 p.m. submission deadline. Unlike our partners that left, we had never done a competition before and so we didn't really know what to expect moving forward. We just decided to quit and not even turn in the proposal. Our professor was understandably upset when we told him what happened. We found out later that the two teammates who left, departed the hotel and went home. It was probably one of the worst experiences of my entire time in college. The professors told us it would be such a great experience and that we would get so much out of it. Other classmates that had done competitions told me the same thing. I don't know what happened with our team, but it was not the experience I was told it would be, nor what I expected.

How Might This Be Changed to Provide a TALE?

This team had entered the same competition as Professor Sey's team, so it had all the potential to produce a great experience. Let's consider some of the factors at play that had they been reversed would have created greater likelihood for a TALE:

- The team was formed as a collaborative team between two universities with a coach from each university. Because of the distance between the schools, the team did not have many opportunities to work together on relevant problems prior to the competition. Thus, positive dynamics had not been developed between team members prior to the competition.
- One of the schools involved had done a substantial number of competitions while the other school had not. Thus, one group of students seemed to understand both expectations and the actual competition; one did not.
- Without previous experience of solving open-ended problems as a team where they could get regular feedback and reflect on their learning, the group quickly focused on their failures as opposed to their successes.

In the end, not all was lost. The two faculty coaches took the opportunity to have the students reflect on the experience and what caused the breakdown and ultimate failure of the team. Through this process, they were able to instill some connection with the importance of team unity and trust in the success of any team endeavor. As well, they were able to show the students how this type of situation happens in the real world and how they can use this experience to better themselves in the future.

Can Competitions Be Done with Limited Resources?

Competitions can be organized in many different shapes and sizes, from two students in a single classroom activity, to a complex organized event held in a large coliseum comprising thousands of students, industry professionals, and sponsors. Achieving benefits doesn't require an elaborate experience that lasts months, includes student travel, and involves other schools. Many of us don't have the resources for that kind of experience, but competitions can be operated at various levels (Figure 11.3):

ITEM	BASIC	ELEVATED	CRITICAL LIFT
JUDGES	Faculty judge(s)	Industry judge(s)	Industry judge(s)
PARTICIPANTS	One class section	Multiple sections and/ or disciplines within a university	Multiple universities (or other way to generate diversity)
DURATION	Short duration (1 class period)	Medium duration (in and out of class work, 2-3 class periods)	Long duration (in and out of class work, multiple days)
AWARD	Pride/Class Recognition	Trophy/Department Recognition	Cash Prize/University Recognition
ATTIRE	No requirement	Dress up (business casual)	Dress up (business formal)
LOCATION	Classroom	Conference Room	Neutral site (e.g. hotel conference center)
RESOURCES REQUIRED	Development time; no additional costs	Increased time; increased cost	Significant time; sponsorship

Figure 11.3 Levels of Competition Execution.

Conclusion

Competitions provide an incredible vehicle to create TALEs for students. While we've suggested a diverse array of ways to execute a competition, there are several items critical to a quality experience in the AEC environment. These include the following:

- The competition involves real-world problems like the work of AEC professionals.
- Competition problems include open-ended inquiry and require teams to engage in discourse to solve problems.
- Students present their findings.
- Formal, timely feedback is provided from the experts judging the event.

With the elements above, competitions can make learning fun and create a perfect environment for a TALE.

Reference

Baden, W. (2006). Play and learning in the classroom: Confessions of a pez dispenser. *The Classical Journal*, *102*(1), 69–71.

12 Studio and Lab Classes

How do you "up the game" in classes that are inherently application-based? What techniques can be incorporated to make these types of classes transcendent for students?

The ability to apply theories and concepts is a key aspect to increasing a student's level of learning. This is especially true in application-intensive disciplines, such as those in the built environment. The physical space of the traditional lecture style classroom isn't ideal for practicing knowledge application. Lab and studio classes offer opportunities for this to occur. The inherent makeup of these class types might lead you to the belief that Transformational Active Learning Experiences (TALEs) are naturally baked right into the class. This is far from the case, and, in fact, it can be more common than you might think for these classes to be less productive for learners. However, given appropriate attention and planning, these class types provide an excellent opportunity for TALEs.

The Basics

Studios and labs are intended for learning by doing. Often, they begin with a small lecture or demonstration followed by student efforts to develop results based on what was just taught. These classes are designed to provide necessary time for focused attention on application.

Lab classrooms are set up with the appropriate equipment and space to execute the necessary learning activities. Oftentimes, each lab class has a very specific activity that is required to be completed during that class. Lab spaces are typically "hot-desks" that provide a temporary space specifically for a student or group of students to complete the scheduled activity.

Studio classes may be something less familiar to you unless you're in a design discipline. They offer an opportunity for applied learning but have some unique characteristics that make them quite different from labs. Contact hours in a studio format are three to four hours per class and usually meet two to three times per week. Studio classrooms are often populated with "cold-desks" – workstations

DOI: 10.1201/9781003106029-12

Figure 12.1 Example Student/Instructor Engagements in Studios. (Photos courtesy of Auburn University College of Architecture, Design and Construction.)

assigned to each student that become their personal workspace they have access to for the entire semester.

The atmosphere and cadence of a studio class can also be much different from a lab. Many class periods may not have a specific activity to be completed. Instead, students spend time working on completing larger, longer-term projects. During these class periods, the instructor will interact with the students in various ways (Figure 12.1). This could be talking with each student individually at their desk about their project (desk critiques) or meeting with small groups of students to informally present and discuss what each of them is doing with their project. Students also conduct formal presentations of their work to a panel of reviewers that may include other faculty within or outside the department and industry professionals (juried review).

What Are the Benefits?

Some of the benefits to these class types are obvious – application of knowledge, opportunity to interact with classmates, and structured instructor feedback. Both class types generally have a lower student/teacher ratio which allows for a more personal learning experience. These types of classes also offer opportunities to repeatedly practice application, increasing learned ability. It likely goes without saying, but another key benefit of these class types is teacher feedback. Labs and studios provide an opportunity for immediate student feedback from the teachers about the work they are doing and what they can do to improve before it has a grade impact. Labs provide focused attention on specific content, skills, or abilities which creates an opportunity to identify specific areas of strength and weakness.

A key aspect of studio classes is for students to practice iteration, and this helps encourage curiosity. The tradition of the studio format focuses on creating a student learning community. With the combination of the increased contact hours and cold-desk approach, students often find themselves spending most of their time in the studio. This approach influences increased collaboration and bonding among students.

Are There Drawbacks?

Labs:

- Often taught by graduate students
- Compressed timeframe for specific activities
- May not allow for repetition/iteration

Studio:

- Often self-paced between major deliverables
- Very time-consuming for students and teachers
- Resource-intensive

Representative Active Learning Strategies for Studio and Lab Classes

There are a variety of other active learning tools that can be used in studios and labs (Figure 12.2):

Representative Active Learning Strategies for Studio and Lab Classes

Peer review/Peer evaluation
Similar to think-pair-share, peer review allows students to participate in low-stakes review of work while engaging in higher-level cognitive skills by analyzing and evaluating others' work.

EXAMPLE: After completing a small lab activity, students swap their work with a classmate and they review each other's work, paying particular attention to correct application of theories and methods. After completing the review, you go over the solution with the class.

Concept maps
Students draw connections between parts of a concept to visually illustrate their understanding. Some examples might include steps in a process or relationship of people in an organization or team.

EXAMPLE: In an engineering survey class, students draw the process for "closing a loop".

Brainstorming
Students work in groups to quickly generate idea responses to a question, topic, or problem. This is a good technique to use at the beginning of a new topic when trying to assess the student's current level of knowledge and understanding.

EXAMPLE: To introduce the concept of Design Charrette's to students in your studio course, you develop an activity where you pose a design question to the students and give them 15 minutes to work in groups of five to formulate multiple solutions which are shared with the group.

Figure 12.2 Selected Active Learning Strategies That May Work Well in Studios and Labs.

The TALE

It was the end of the semester; Addison and her team were preparing to give their final project presentation for their third-year design studio class. Although it was common for nerves to be on edge during final studio presentations, this time they were exceptionally high. It was the first time they gave a presentation to a real-world client. The student teams were presenting design proposals for a local fire station, and the goal was for the fire department to select one of the student designs that they would use as a base model for their new building. Students in the class had invested the entire semester working toward this event. For Addison and her group, this was the culminating event.

For the four professors teaching the studio, the fire station project was not an accident. The professors contacted architects about their class collaboration and found the local fire station opportunity. Working with the architect and fire department, the professors built the experience using an actual project with an actual client – all within the context of the course learning objectives.

To the students' surprise on the first day of studio, the professors introduced the firefighters and architects as the students' clients. Early in the semester, the students took a tour of a different fire station. The firefighters walked the students through the station, explaining the different aspects of the building, identifying key necessities, likes, and dislikes. At the end of the tour, the students had the opportunity to ask the firefighters questions to better understand what their goals for the new building were. Similarly, the architect would visit the class almost bi-weekly to review the student's work, discuss how they execute their processes, and provide guidance.

At the midterm, the student teams met at the architect's office to present their progress to the team and professors for feedback and suggestions moving forward. The professors set up the expectation for this meeting a week earlier by having the architect visit the class and give a presentation on design progress meetings with a client. Addison recalls how that class raised everyone's motivation to do their best. It must have worked because the architect and fire department commented how impressed they were after the meetings with the teams:

> "They've come up with some amazing ideas."
> "If we had to stop right now, we would already have some great ideas to work with."

Following the midterm reviews, Addison remembers how the constructive feedback from the team was instrumental in helping them move forward and gave the students a real sense of what it would be like in the real world. Leading up to the final presentation, the professors organized guest presentations from the architects' consultants to help the students with technical questions at key points through the design. Addison felt their team was well prepared for the final presentation. But now that the time had come, the realization of what they were about to do became very real. The engine bay of the fire station where they were presenting was filled with about 30 fire fighters, the architect's staff, their professors and classmates, members of the community, and a local newspaper, all waiting to watch them.

A few hours later, they were still in the fire station, but instead of nerves and apprehension, the mood was jubilance. Gathered at tables after a celebration lunch, the students listened to the fire chief and architect speak about the great work of all the students and the wonderful experience they had on the project. They ended by announcing the selected team design, and although Addison's team wasn't selected, she didn't care; the experience was amazing, and just getting to participate created an immense sense of pride.

Connecting HIPs and Moments in the TALE

While some of the key aspects of Addison's experience are likely obvious, Figure 12.3 highlights why her experience was so profound.

Elevation

Having the firefighters and architect unexpectedly come to the classroom on day one to help introduce the semester project broke the script. The continued interaction with clients and the final presentation raised the stakes.

Insight

The field trip to the fire department and the midterm presentation at the architect's office provided an opportunity for the students to have real-world experiences that were set at an appropriately high expectation. It also allowed them to demonstrate their knowledge and skills, providing an opportunity to trip over the truth and stretch for insight.

Pride

Numerous aspects of the experience created an opportunity for pride:

- A semester spent working on the project for an outside client
- Multiple presentations publicly demonstrating their knowledge and skills
- A recognition event by the fire chief and architect

Connection

Having the students work in teams helped strengthen the ability for them to share in struggles and victories. And the numerous guests that visited the class over the semester allowed greater connection through frequent opportunities for feedback.

TALE Strength Summary

Where do the HIPs and Moments intersect in your proposed plan?

Elements of HIPs	Components of MOMENTS			
	ELEVATION	INSIGHT	PRIDE	CONNECTION
What will you do to **set the expectations** so they are challenging for the students but not discouraging to the point they give up?	X	X	X	
How will the experience be set up to require a **significant investment of time and effort** by the students?			X	X
How can the experience involve the students having **interaction with faculty and peers**?	X	X	X	X
What can be done to have students **interact with others** that they wouldn't normally interact with in the traditional class experience?	X	X	X	X
What aspects of the experience will **provide** students with **frequent, timely, and constructive feedback**?		X		X
How can you incorporate **periodic opportunities within the experience for students to reflect** and integrate what they've learned?		X		
How can you make the experience feel more **real-world** for the students instead of a ficticious learning exercise?	X	X	X	X
What can you do to have the **students demonstrate their knowledge and skills** to others outside the class (public display)?	X	X	X	X

X *Indicates an area where HIP element and Moment align*

Figure 12.3 Intersection of HIP Elements and Components of Moments for the TALE.

Crash and Burn

Professor Brady was looking for a way to actively engage his construction management students in studying the concept of materials in concrete mix designs and the role of reinforcement in concrete structures. He developed a lab where his students would work with architecture students in collaborative teams to design and build a concrete bench. Professor Brady had reached out to Dr. Golden (the architecture professor) at the beginning of the semester to set up the collaboration. Professor Brady assured Dr. Golden he would not have to do anything other than send his students to the class on the day the assignment was introduced, and that he would handle the rest. Dr. Golden obliged but warned Brady that his students also had a significant course load and that their grade wouldn't be impacted at all by the collaboration.

On the day the assignment was introduced, the combined class had the opportunity to meet each other, form teams, learn about the scope of the project, and start work. Professor Brady explained the scope of the project and that they would have one week to complete and present their concrete bench projects. During the introduction, he also informed them that they would not meet for the regularly scheduled class times the rest of the week so that the students could use that time to work on the project.

On presentation day, Professor Brady was filled with anticipation to see the student's completed projects. He arrived at the lab classroom about 20 minutes early thinking he would be able to preview some of the student's work before beginning presentations, but upon entering the classroom he was dismayed at what he saw. No students were there; the lab was a mess with materials the students had clearly been using to build their projects; and it appeared only two of the seven projects were complete. Professor Brady couldn't understand what was going on and could only wonder if the students thought the completion day was later in the week. As disgruntled students started trickling in about five minutes before class, it became very clear to Prof. Brady that the students encountered numerous issues including team conflicts, lack of materials, failing concrete casts, designs too complex to build, procrastination, and poor collaboration. Professor Brady was now left with deciding how to address the situation rather than being able to give students feedback on their projects.

How Might This Be Changed to Provide a TALE?

Professor Brady had a sound idea that included high expectations, interaction with peers and others outside their discipline, public display of knowledge, components of elevation, insight, and connection. But there were critical aspects he missed that could have created a dramatically different outcome:

- Getting Dr. Golden more involved (connected and engaged) could have helped with feedback.
- Making the project timeline longer than a week would have allowed more time for the students to learn, develop insight, build pride, and reflect and revise their work.

- Incorporating an outside party such as local businesses that would use the completed projects would increase the performance expectation, elevation, and pride.
- Opportunity for intermediate feedback from him could have identified issues early associated with improper mix-designs, designs too complex to build, and team issues.

Although the stakes were high for the CM students, there weren't any stakes for the architecture students. This created unequal motivation. Often this results in one group of students attempting to salvage a project with limited or no involvement from the other, collaborating group. Equalizing the stakes often yields better results for collaborative activities.

Conclusion

Studio and lab classes offer great opportunities to build TALEs. The extra time built into these classes for students to practice application incorporates many details of a TALE, just as we saw in Addison's experience. However, including aspects of a TALE doesn't guarantee success, just as we saw with Professor Brady's experience. When trying to build a TALE, careful consideration needs to be given to not just the factors included, but especially those left out and their related impacts.

13 Math-Intensive Classes

> When you teach math-based principles, in what ways do you teach the math separate from the practical application of task? In what ways do you integrate the math with the design or engineering principle being considered?

Over the history of the built environment, math has served architects, engineers, and constructors well. Designers and builders must think mathematically and use math concepts to design and build the projects they seek to create. As times have evolved and technology has moved to the forefront of professional practice, many of the calculations previously done by hand are now done by a computer. An interesting tension among students is created because industry uses computers for calculating while faculty teaching foundational knowledge tend to rely on hand calculations. This amplifies the need to make math-intensive classes relevant to professional practice while also conveying the time-tested skills of problem definition, problem solving, inductive and deductive reasoning, information ordering, and communication. Beyond making it relevant, we also recognize that some students find math difficult and see it as a barrier instead of a problem-solving skill.

The Basics

Several subject areas in the built environment require the significant use of mathematics. These could include an introduction to physics, calculus, or an applied discipline like statics, dynamics, environmental systems, electrical circuits, estimating, or thermodynamics. Many of these classes have a foundation based on mathematics, and as a result appear to students as more computational than practical. Classes often involve significant calculations followed by problem-based homework often requiring significant investment of time and effort by students to master intended concepts.

With 21st-century technology, students often misperceive that they don't need foundational math competencies because technology "does the math." However, math competencies used by built environment professionals must complement the work of computers. Modeling and applications are key within this environment. Students must understand the math that supports the work of computers.

DOI: 10.1201/9781003106029-13

In other words, a conceptual understanding of the key principles is critical. Mathematics for built environment professions is best learned by presenting problems in realistic contexts.

What Are the Benefits?

The history of the built environment provides strong support that math-intensive classes are essential to the critical thinking skills so important in the AEC industry. The process of defining the problem, stating the given information, applying relevant formulas, redefining the problem, solving the problem, and implementing a solution has proven a very effective strategy for developing great designers and builders. Once a solution has been implemented, further value is gained from evaluating the solution and reimaging the problem yet again.

Are There Drawbacks?

Many students struggle with math-intensive classes. Much of this comes from a predisposition of poor performance and/or a perception that they will not use it in professional practice. In many cases, faculty have not helped this process, often separating the math-based foundation of the material from the applied, contextual knowledge. Without domain-specific examples, students are left feeling disconnected, and the "use-value" is lost.

Math-based classes are a challenge for program retention. Students often cite a lack of confidence in math as opposed to mathematical capability as a key reason for leaving design and build professions (Kuo, 2016). One report noted that 50% of students that enroll in college Algebra do not make a C or higher (Saxe & Braddy, 2015). Some suggest that math-based classes are the primary barrier to achieving a degree within the built environment.

Representative Active Learning Strategies for Math-Intensive Classes

Active learning strategies represent an opportunity to mitigate some of the drawbacks associated with math-intensive classes. By engaging students at a deeper level and providing opportunities for reflection and discussion through specific active learning strategies, students' work may improve (Figure 13.1).

The TALE

Professor Rohn was well loved by students in the mechanical engineering department. His introductory class on heat transfer had significant math components, but students appreciated the connection he made between the required mathematics and practical applications in the middle of Fourier's Law, Boltzmann's Law, and differential equations. Professor Rohn focused on providing strong connections with practical applications on two or three major topics during the semester.

Representative Active Learning Strategies for Math-Intensive Classes

Clarification pauses

After presenting a key point or concept, pause and give the students time to think about the issue, then ask if anyone needs anything clarified. One way to do this is to ask students to review their notes during the pause.

EXAMPLE: In an architectural class exploring the allowable number of people in a building per code, the instructor calculated how many doors, toilets, and parking spaces would be required for each of the user groups. After each mathematical calculation, he paused and requested students review their notes and ask for clarification.

Cooperative groups/Informal

Consider forming groups of 3-4 students to use throughout the semester. During a lecture, pose a question for the groups to discuss between themselves. While groups are discussing, the teacher will walk the room stopping at several groups to keep the groups focused on purpose. Groups can be asked to "report out" to the class on what they found.

EXAMPLE: In a course on deep foundations, one class considers side friction and end bearing piles. The instructor develops three questions for groups to consider: 1) For a 12" square displacement pile that is 30' in length (with some typical friction and end bearing values), is the side friction or end bearing more dominant? 2) How might this effect what size of pile you select? 3) What would change with a non-displacement pile?

Interactive learning

Find time during the lecture to include an activity that allows students to work with the material being covered. This could include the students solving a problem posed, exploring graphics, or connecting the material to professional practice.

EXAMPLE: Find time during the lecture to include an activity that allows students to work with the material being covered. This could include the students solving a problem posed, exploring graphics, or connecting the material to professional practice.

Figure 13.1 Selected Active Learning Strategies That May Work Well in Math-Intensive Classes.

One of those points of connection was on heat transfer through buildings. In general, the amount of heat that is transferred through a wall of a home is dependent on the characteristics of the material and its thickness. Heat is transferred by radiation, convection, or conduction. Thermal resistance is a measure used to represent the ability of a set of materials to resist flow of heat through the wall.

Professor Rohn secured a thermal imaging camera through a departmental grant, and he worked with industry partners to build a testing system where he could measure heat flow through a variety of wall systems. In short, a box was constructed with an internal heater. An opening in the exterior wall of the box provided a spot to insert a variety of small wall systems. In one case, a window section could be tested, while in another an EIFS system could be tested. The thermal imaging camera was then used to determine heat change through the wall system. Such a system provided a visual and tangible application of the heat transfer topics discussed in class.

Professor Rohn began his discussions of heat transfer simply by bringing a cup of coffee to the first class and asked students to describe what happened over time to the hot coffee. This simple example set the expectation for the students that the topics they would discuss were relevant to their lives. Professor Rohn mixed foundational mathematical concepts with applied heat transfer issues primarily associated with exterior wall construction. In early assignments, students explored heat transfer through a single material. More advanced discussions led them to explore heat transfer through multiple materials. The problem-solving was more complex as the material advanced, but each key topic was bolstered by assigned problems that required students to apply what they were learning in class. Professor Rohn graded those assignments quickly providing prompt feedback on problems worked. Students then participated in labs using the "Rohn Box" to predict and measure heat transfer through a variety of wall constructs. When predicted values differed from actual measurements, students were challenged to reflect on the work and offer possible reasons for the differences.

Professor Rohn culminated the lessons with a project where small student groups had to build their own wall systems that would fit in the box using multiple materials. Industry partners were engaged to support student wall construction. Students would predict heat transfer using mathematical formulas and then test their wall systems in the "Rohn Box". Professor Rohn would then provide feedback and suggest why mathematical models may not have matched actual applications. Students loved the tangible connection to the real world and appreciated multiple applications of real-life heat transfer explored throughout the semester.

Connecting HIPs and Moments in the TALE

Let's see if we can pinpoint some of the factors that created such strong educational experiences (Figure 13.2).

Elevation

Professor Rohn used elevation to engage the students deeper at every opportunity. From starting class with a steaming cup of coffee to the development of the "Rohn Box," almost all class events were developed to illustrate clearly how math was a tool to think about real-world phenomena.

Insight

By exploring heat transfer in a variety of conditions, students were challenged to apply the mathematical concepts and think deeply when models either matched

TALE Strength Summary

Where do the HIPs and Moments intersect in your proposed plan?

Elements of HIPs	Components of MOMENTS: ELEVATION	INSIGHT	PRIDE	CONNECTION
What will you do to **set the expectations** so they are challenging for the students but not discouraging to the point they give up?	X			
How will the experience be set up to require a **significant investment of time and effort** by the students?	X			
How can the experience involve the students having **interaction with faculty and peers**?	X	X		
What can be done to have students **interact with others** that they wouldn't normally interact with in the traditional class experience?				
What aspects of the experience will **provide** students with **frequent, timely, and constructive feedback**?	X	X		X
How can you incorporate **periodic opportunities within the experience for students to reflect** and integrate what they've learned?	X	X		X
How can you make the experience feel more **real-world** for the students instead of a ficticious learning exercise?	X	X	X	X
What can you do to have the **students demonstrate their knowledge and skills** to others outside the class (public display)?				

X *Indicates an area where HIP element and Moment align*

Figure 13.2 Intersection of HIP Elements and Components of Moments for the TALE.

or did not match the actual performance of wall systems. Professor Rohn's consistent and prompt feedback added to the students' insight as they were constantly challenged to think differently.

Pride

Pride was realized primarily when students were able to predict and then confirm in practice the heat transfer through various materials.

Connection

Students connected strongly with Professor Rohn's approach, and the discussions fostered among peers as they discussed various aspects of heat transfer fostered additional connections.

Crash and Burn

Professor Goode knew he conveyed a lot of relevant, practical information in his introductory class in electrical engineering. In an effort to provide a theoretical foundation for his students, he spent the first six weeks of the semester reviewing topics from Physics II. Lectures and problem-solving tasks involved fundamentals of electrical charge and flux, electric fields, Gauss' law, electric potential, capacitance, current and resistance, and DC circuits. Professor Goode knew that understanding this information would provide a background for the applied and tangible items covered later in class. But his reviews indicated that students were just not getting the material. One student commented, "I gave up after the first three weeks. All the theory and math really had me wondering why I was in the class."

One student conveyed a particularly troublesome lab she attended early in the class. Shawna felt strongly she had understood the lecture on Ohm's Law for simple parallel circuits. She was excited to attend a lab on parallel and series circuits. She was certain the material would make the topic relevant for her and connect it to applied practice. When she arrived at the lab, she was expecting to see examples of parallel circuits but instead the graduate assistant that teaches the lab presented them with a math lab. Instead of understanding how circuitry was applicable, she was left to calculate resistance in each parallel and find the total resistance. During the lab, the TA also gave her a list of rules to apply which included the following:

- All components share the same voltage.
- Resistances diminish to equal a smaller, total resistance.
- Branch currents add to equal a larger, total current.

Disappointed and confused on how these would apply in practice, Shawna could not connect the relevance of the material to her learning. After three hours in the lab, she left wondering what she had missed and why she had ever been interested in electrical engineering.

How Might This Be Changed to Provide a TALE?

By divorcing the theory and mathematical background from the applied materials, Professor Goode had crashed and burned before even getting to what he saw as the most valuable material of the course.

- Instead of working so hard to repeat the electrical material from Physics II, Professor Goode could start with a set of real-world problems students should be able to solve after completing the class.
- Every new unit in the class could begin with Professor Goode introducing a main problem that the students would be trying to solve. The lecture portion of the class could then focus on the related theory and math associated with solving the problem.
- The TA-taught lab could focus on applying theory to solve the problem. Professor Goode being more involved in the lab classes could also create an opportunity to build a connection between theory and real-world application.

Conclusion

Math-based classes will continue to be critical to educating the next generation of designers and builders. Students will continue to seek great instructors that present concepts from multiple perspectives and employ strong examples and applications from industry to motivate students. Instructors should also set expectations that math-based courses provide an opportunity for students to learn to communicate complex ideas in ways understandable to industry partners, clients, and employers.

References

Kuo, M. (2016, July 22). Low math confidence discourages female students from pursuing STEM disciplines. *Science Magazine*. https://www.sciencemag.org/careers/2016/07/low-math-confidence-discourages-female-students-pursuing-stem-disciplines

Saxe, K., & Braddy, L. (2015). *A common vision for mathematical sciences programs in 2025*. Mathematical Association of America. https://www.maa.org/sites/default/files/pdf/CommonVisionFinal.pdf

14 Tech-Intensive Classes

> Built environment projects use a wide array of technologies. Students who are well-versed in technology are often valued by companies. How could you see constructing a TALE around the application of technology?

Imagine a class where the technology component dominates the curriculum, and outcomes focus on students applying principles and knowledge through the latest technology – this is a tech-intensive class. Nowadays, we take for granted that our students are tech-savvy. But understanding how a tablet works do not necessarily translate to the advanced hardware and software specific to the Architecture, Engineering, and Construction (AEC) industry. In today's industry, students are expected to, at a minimum, have a basic understanding of domain-specific technology.

For faculty, teaching tech is difficult – interoperability issues, constant versioning updates, interface changes, and emergent technologies – not to mention the need to then communicate information to students with varying tech-based comfort levels. Nevertheless, today's industry requires students to understand technology and its application. Even though it is complex, tech-intensive classes can be transitioned from "training" of technology to transformational experiences motivating students to be agents of change within their new companies.

The Basics

Tech-intensive classes are quite variable depending on the content. In architecture, tech is often design software, such as Rhino 3d, Revit, Sketchup, Adobe Suites, V-Ray, and AutoCAD. Civil Engineering students will often learn design software and hardware applications like MATLAB, ArcGIS, HEC-RAS, HEC-HMS, STAAD Pro, and AutoCAD 3D. Construction Management students are often taught various software such as Revit, Procore, Bluebeam, Primavera P6, Microsoft Project, OnScreen Takeoff, and Sketchup. Regardless of the content, a typical tech class will use a tutorial format – a step-by-step demonstration-based approach, followed by an in-class lab, then a more complex out-of-class assignment, and concluding with a feedback loop.

DOI: 10.1201/9781003106029-14

What Are the Benefits?

The benefit of teaching technologies to AEC students is obvious – the industry is tech heavy, and students who are proficient, or at least conversant in the latest tech, are in a valued category of potential employees. In some companies, recent college graduates and those most comfortable with technology will drive innovation. This could advance the graduate's positioning within a company quickly. In addition, learning advanced software reinforces the fundamentals of "every day" software. For example, proficiently writing equations in Python could help clean up formulas in Excel. Today, educators must not look at knowledge and technology as mutually exclusive; the two are inextricably linked. While the understanding and derivation of complex topics is important, the application often relies on modern technology.

Are There Drawbacks?

Some drawbacks to teaching tech are unavoidable – troubleshooting hardware and software, up-front cost, licensure renewals, seat quantity, and interoperability with legacy software are all issues that are inevitable. Dealing with these issues is just the cost of doing business. That said, other drawbacks, such as complexity of pedagogy, varying comfort levels among students, and poor retention can be mitigated with the right approach. Students may also miss the relevance of the tech they are learning to their future career. Students can also be distracted from other intended learning outcomes by focusing more on learning how to use the software than how to apply it.

Representative Active Learning Strategies for Tech-Intensive Classes

As we have mentioned, teaching technology is much more than just "training" of software and hardware. Incorporating some of the strategies in Figure 14.1 can help facilitate the elements of critical thinking, problem-solving, and inquisitiveness that can elevate teaching technology from "training" to deeper learning.

The TALE

Scheduling a complex construction project is no walk in the park – it involves coordinating dozens of subcontractors, hundreds of deliveries, resource availability, delays, changes, and unforeseen conditions for the life of the project, all based on robust software that has a significant learning curve. It is a job for problem-solvers and thinkers. This is where Colton thrives. But three years ago, as a senior in college, he was not excited about this class. In fact, like most of his colleagues, he knew that the class was difficult and was overwhelmed with the volume of information and, in particular, the software that aids project schedulers. After a few weeks, the class got into the complexities of project scheduling and software, and

Representative Active Learning Strategies for Tech-Intensive Classes

Retrieval practice

Pause for an appropriate amount of time every 15 minutes, having students write or execute everything they can remember from the preceding class segment. Encourage questions.

EXAMPLE: When teaching robotics, each time a new payload is introduced and explained, ask the students to produce an application document. This document will note the previously learned payloads and what its use is for the robot. This formative review helps with long-term retention.

Demonstrations

This approach asks students to predict an outcome for a demonstration that they will witness. Once the demonstration is completed, students will discuss what they observed, anything that was different from what they predicted, and suggest what may have caused the outcome to be different than what they predicted.

EXAMPLE: While teaching a visual programming software, such as Dynamo, create a string and ask the students to predict what that line of code will do to the Revit model. Once the class finds consensus, run the code. Close the discussion loop by asking the group if the result differed from their expectations.

Hands-on tech

Students use technology such as software, visualization, or communication devices to enhance understanding of concepts being studied.

EXAMPLE: To understand hydrology, engineering students interact with an Augmented Reality Sandbox. Moving the sand around changes the topography of the land, and the built-in AI of the software adjusts the water runoff accordingly.

Figure 14.1 Selected Active Learning Strategies That May Work Well in Tech-Intensive Classes.

to his surprise, something in him clicked. He started to see the logic in the chaos. Part of this epiphany is related to Colton's problem-solving nature, but the other part was how the class was taught.

Dr. Ryan had years of experience on construction sites before making her way back to academia. She understood the intricacies of construction scheduling, as well as the complexities of teaching software to students. Her approach to teaching tech relied on a principle of growth. She never expected her students to create schedules to match industry standards, especially early on; rather, she expected the students to vastly improve over the course of the semester – and her grading reflected that. To Dr. Ryan, teaching tech was all about exposure to numerous examples, timely feedback, and application. The students must interact with the tech over an extended period. No tech can be learned and applied without a time investment; there must be growth over time, and the feedback must come in a form that is conducive to learning. The software was integral to how scheduling would be mastered by the student and implemented in practice by the professional.

Colton recalls how the software was first introduced in the class – presenting how it is used in today's industry and the current software in comparison to the legacy approach. This set the stage for the application and relevance for the students. The tech was taught in a single class period, with Dr. Ryan executing the steps of the software with the students mirroring each step on their own devices. This was a slow process due to the varying levels of understanding in the room, but allowed students to ask questions, get help from peers, and clarify reasoning. Following the class, a more concise video of the tutorial was posted to the class hub, walking through the exact same steps but without the interruptions that are present in class. This video was a reference tool for students but became a major guide when the students executed their schedule projects. Over the semester, the students created three schedules for three different, real-world projects in small groups. These three schedule submissions were due around weeks 7, 11, and 15. The longitudinal nature of these assignments was massively important to cognition and retention, not only to the technical nature of building a schedule but also to understand the industry standard software being used to building the schedules.

Perhaps the most important steps Dr. Ryan took in the class were the feedback loops. Each schedule had an intermediate and final feedback loop. Halfway between giving the schedule assignment and its due date, Dr. Ryan would hold "workdays." These check-ins gave students one-on-one time with Dr. Ryan to answer questions about the schedule and, often, the software. The final feedback loop came in the form of "grade day" which was a 30-minute session with each scheduling group to walk through their schedule. During these sessions, Dr. Ryan would review logic, flow, presentation, and ask clarifying questions as to how the students handled the software. After 25 minutes of grade day, the group would sit down and review the feedback with Dr. Ryan. Before leaving their grade day, the students knew their grade and had tangible, actionable direction.

Colton looks back on this class fondly. Like his colleagues in the class, he struggled with the complex scheduling software, but the approach Dr. Ryan took to teaching gave plenty of opportunities to learn through guidance and metacognition. The guided tutorial, the reference material, the intermediate check-ins, the grade day reviews, and the longitudinal nature of learning, all lead to his expertise with scheduling and the software he uses each day in professional practice.

Connecting HIPs and Moments in the TALE

Figure 14.2 summarizes the TALE experienced by Colton.

Elevation

The grade days in the TALE require the students to become proficient enough in the tech to present their schedules to a faculty member. Grade day elevates learning the tech by "raising the stakes" through high expectations and asks students to demonstrate their knowledge by "boosting the sensory appeal."

TALE Strength Summary

Where do the HIPs and Moments intersect in your proposed plan?

Elements of HIPs	Components of MOMENTS: ELEVATION	INSIGHT	PRIDE	CONNECTION
What will you do to **set the expectations** so they are challenging for the students but not discouraging to the point they give up?	X		X	
How will the experience be set up to require a **significant investment of time and effort** by the students?			X	
How can the experience involve the students having **interaction with faculty and peers**?		X		
What can be done to have students **interact with others** that they wouldn't normally interact with in the traditional class experience?			X	X
What aspects of the experience will **provide** students with **frequent, timely, and constructive feedback**?		X		
How can you incorporate **periodic opportunities within the experience for students to reflect** and integrate what they've learned?				X
How can you make the experience feel more **real-world** for the students instead of a ficticious learning exercise?				X
What can you do to have the **students demonstrate their knowledge and skills** to others outside the class (public display)?	X	X	X	

X *Indicates an area where HIP element and Moment align*

Figure 14.2 Intersection of HIP Elements and Components of Moments for the TALE.

Insight

Interaction with the faculty and peers, and the timely feedback through the workdays and grade days, requires students to "stretch for insight" through feedback that promotes growth. Without these interactions and frequent feedback, the application of the tech, as well as the retention of how to use the tech, would likely be significantly lowered.

Pride

Through peer interactions, the elevated expectations, the significant effort put into creating a complex construction schedule, and the resulting grade day presentation, students often achieve significant bouts of pride during the process. Within the small groups, the collective understanding of the tech helps the students develop schedules while mitigating issues. The final product presented on grade day is the result of a great deal of collaborative troubleshooting, and the recognition of peers within grade day is palpable.

Connection

Learning and applying the complex scheduling software in a small group, to real-world projects, creates a "sharing in purposeful struggle" and a "connection to meaning." In addition, the grade day requires students to be reflective in the process and "share in the accomplishment" of applying complex technology.

Crash and Burn

Professor Thomas is about to begin class and is excited to teach her students the Revit software they will use for their studio design assignment. The software is widely used in the AEC industry, and Professor Thomas knows this will be an opportunity for them to build competencies they will use in professional practice. The software is complex, but Professor Thomas has taken a couple of training courses and is quite confident in the task ahead – recognizing she is not an expert. She planned to spend one class period demonstrating the basics of the software and has provided numerous Revit tutorials on the course management page for students to access as needed.

Before starting the class, some students informed her they could not get the software loaded onto their computers, and some were not able to access the full capabilities of the software. After spending the first 15 minutes of class trying to resolve the students' issues with little success, Professor Thomas directed them to the IT department. In the meantime, students could sit with a classmate and follow along while she went through the tutorial.

As Professor Thomas went through the tutorial, students frequently stopped her to repeat the steps they missed. This was frustrating because Professor Thomas had only planned one class for the tutorial, and she needed to get through it so they could focus on design topics she had planned for other parts of the assignment. Adding to this frustration, her screen looked different than theirs, and they lacked the same options on their computer she had. It seemed Professor Thomas's software hadn't been updated to match the students' version. Concerned about finishing the tutorial, she told the students not to worry about it because the tutorial videos she posted should clarify any confusion over the next week as they finished the assignment.

Professor Thomas was able (with the help of the IT department) to get the software working correctly on all students' computers. As the week progressed, students asked questions about using the software – many of which Professor Thomas was unsure how to answer but assisted to the best of her ability. She also began noticing many students' designs were not progressing. Students were frustrated, and Professor Thomas realized it as the week developed. It became clear to Professor Thomas that poor implementation of the software was more of a distraction than a benefit to the learning process, and the failed effort caused students to lose trust in her as their teacher.

How Might This Be Changed to Provide a TALE?

Teaching tech is difficult, and the crash and burn Professor Thomas experienced is not unique. However, there are strategies to mitigate the issues she encountered and turn the experience into something powerful:

- Incorporating an outside professional with expertise in the software could elevate the experience by providing interactions with different people in real-world settings and remediate issues with a lack of expertise.
- Providing more time, over multiple milestones, for the students to incorporate the application of the software would allow them opportunity to receive frequent feedback, reflect on what they are learning and realize insight into how the software is applicable to their profession.
- Having the students work in collaborative groups so they have an opportunity to develop connections and learn from each other would enhance meaningful interaction.
- Incorporating opportunities for students to present challenges they are having and insights they have discovered about the software would allow them to develop pride through shared experiences and public display of knowledge.

Conclusion

Regardless of the discipline, the need to teach emergent technologies is inevitable to keep students at the forefront of advancements in the profession. In each class, there will be various levels of comfort and ability with technology. Even with significant planning and preparation, a crash and burn can still occur. Tech, especially emergent tech, is prone to failure, often at no fault of the user. Instructors who teach technology should be vigilant in preparation, but just go with the flow and do not be afraid to fail in front of the students. In most cases, they appreciate and connect with you more because of your willingness to be vulnerable in front of them.

15 Afterword

Every student deserves multiple Transformational Active Learning Experiences (TALEs) within their college curriculum, and universities that find ways to create these will have a competitive advantage over others in the marketplace. Deeply impactful, lasting educational experiences are both touchstones of a college career and an effective mechanism for cognition and retention. It is our hope that the process and examples provided in this book inspire creativity and enthusiasm, while allowing you to be efficient in the creation of your own TALE.

The story of this book is a simple one. Three friends and colleagues passionate about teaching and learning, with a significant amount of teaching and research of teaching under their collective belts, decided to write a book. Reflecting on their college experiences, successful (and not so successful) TALEs they have created, and discussions with colleagues on their experiences led to the creation of this manuscript. The process presented provides a recipe to fabricate a TALE in almost any class imaginable while allowing faculty to "start small" and build over time. And believe it or not, the creation of this book deepened our friendship and created our own TALE.

DOI: 10.1201/9781003106029-15

Index

Note: *Italic* page numbers refer to figures.

For Product Safety Concerns and Information please contact our EU representative GPSR@taylorandfrancis.com
Taylor & Francis Verlag GmbH, Kaufingerstraße 24, 80331 München, Germany

www.ingramcontent.com/pod-product-compliance
Ingram Content Group UK Ltd.
Pitfield, Milton Keynes, MK11 3LW, UK
UKHW011019180926
475701UK00004B/834

9780367616786